T0176124

Modeling Neural Circuits Made Simple with Python

Computational Neuroscience

Terence J. Sejnowski and Tomaso A. Poggio, editors

For a complete list of books in this series, see the back of the book and
https://mitpress.mit.edu/books/series/computational-neuroscience

Modeling Neural Circuits Made Simple with Python

Robert Rosenbaum

The MIT Press
Cambridge, Massachusetts
London, England

This book was set in Times New Roman by Westchester Publishing Services. Printed and bound in the United States of America.

Library of Congress Cataloging-in-Publication Data

Names: Rosenbaum, Robert (Researcher in computational neuroscience), author.
Title: Modeling neural circuits made simple with Python / Robert Rosenbaum.
Description: Cambridge, Massachusetts : The MIT Press, [2024] | Series: Computational
    neuroscience series | Includes bibliographical references and index.
Identifiers: LCCN 2023026888 (print) | LCCN 2023026889 (ebook) | ISBN 9780262548083 (print) |
    ISBN 9780262378758 (epub) | ISBN 9780262378741 (pdf)
Subjects: LCSH: Neural circuitry—Mathematical models. | Neural circuitry—Computer simulation. |
    Neural networks (Neurobiology)—Mathematical models. | Neural networks (Neurobiology)—
    Computer simulation. | Python (Computer program language)
Classification: LCC QP363.3 .R67 2024 (print) | LCC QP363.3 (ebook) |
    DDC 573.8/5–dc23/eng/20230927
LC record available at https://lccn.loc.gov/2023026888
LC ebook record available at https://lccn.loc.gov/2023026889

10 9 8 7 6 5 4 3 2 1

All models are wrong, but some are useful.

—George Box

Everything should be made as simple as possible, but no simpler.

—paraphrased from Albert Einstein

Contents

List of Figures

Preface

This book assumes a basic background in calculus (derivatives and integrals), linear algebra (matrix products and eigenvalues), and probability or statistics (expectations and variance). All other mathematical background is covered in appendix A. Before each paragraph that introduces a new mathematical concept, there is a red box containing a reference to the relevant section in appendix A:

> References to appendix A of mathematical background look like this.

If you need to learn or review that concept, follow the reference before going on. Otherwise, you can ignore the reference and continue reading.

When using this textbook for a course, each section in appendix A can be covered in a lecture as a detour from the main material, or it can be covered in an optional *recorded* lecture if students are expected to have learned the topic before.

In addition to its use as a course textbook, this book is well suited to independent reading. It might particularly appeal to readers with a background in machine learning who would like to understand the relationship between artificial neural networks and common biological neural network models.

Figures in the book are accompanied by Python notebooks containing the code needed to reproduce the figure. See the book's website for links to the Python notebooks. The file name of each Python notebook is referenced in the associated figure caption. If you are not familiar with Python or NumPy, or if you just need a review, see the notebook `PythonIntro.ipynb` for a brief introduction to Python programming.

The main text of this book was whittled down to a minimal thread of material needed to build a backbone for modeling neural circuits. This approach assures that the book can be covered in a one-semester course without rushing, and it also makes the book amenable to self-learning. Some important topics and models that were omitted from the main text are covered in appendix B. References to appendix B appear within their relevant context throughout the book:

The decision to omit the details of ion channel dynamics and the Hodgkin-Huxley (HH) model from the main text was not made lightly, but they are not needed for understanding the remainder of the book, and their omission allows an instructor to spend more time on other topics. If you are teaching a course with this textbook and want to include these topics, then you should take a detour to appendix B.1 after covering section 1.1 of chapter 1.

Some sections in this book were inspired by similar sections in the two excellent textbooks, *Theoretical Neuroscience* (Dayan and Abbott 2001) and *Neuronal Dynamics* (Gerstner et al. 2014), which can serve as supplements to this book. For a very nice exposition on the history of computational neuroscience and some of its fundamental concepts, read Grace Lindsay's popular science book, *Models of the Mind* (Lindsay 2021).

Acknowledgments

I would like to thank Adam Kohn, Matthew Smith, Micheal Okun, and Ilan Lampl for providing the data used in chapter 2, appendix B.4, and exercise 4.2.1. I would also like to thank David Toth, Vicky Zhu, and all of the other students whose feedback and comments helped shape the contents of this book. Work on this book was supported in part by funding from the US National Science Foundation CAREER Award number DMS-1654268.

1

Modeling Single Neurons

Neurons are arguably the most important type of cell in the nervous system. In this chapter, we will develop mathematical and computational models of neurons, focusing primarily on neurons in the *cerebral cortex*, which is widely viewed as the central processing area of the mammalian brain. The cortex is a sheet of neural tissue with multiple layers, folded up to form the outermost part of the brain. There is a large diversity of neuron types with a variety of properties, but the prototypical cortical neuron is in the ballpark of 10μm (micrometers) in size and composed of three parts (figure 1.1a): the dendrites, soma, and axon. Dendrites are tree-like structures on which neurons receive input from other neurons at connections called *synapses*. The soma is the cell body where these inputs are integrated. The neuron's response to its inputs propagates down the axon, where it can be communicated to other neurons.

Neurons maintain a negative electrical potential across their membrane, meaning that the ratio of negatively to positively charged ions is greater inside the cell than outside it. Specifically, the potential across the neuron's membrane is usually around -70 mV (millivolts). This electrical potential is called the neuron's *membrane potential*, which we denote as V. The membrane potential is modulated when positively or negatively charged ions flow through *ion channels* in the membrane.

When a neuron's membrane potential reaches a threshold around $V \approx -55$ mV, the opening and closing of different ion channels creates an *action potential* or *spike*, which is a deviation of V to around 0–10 mV that lasts about 1–2 ms (figure 1.1b). Spikes propagate down the neuron's axon, where they activate synapses. The synapses open ion channels on the postsynaptic neuron's membrane, causing a brief pulse of current (figure 1.1b). In this chapter, we construct mathematical and computational models of these dynamics, which are sketched in figure 1.1.

1.1 The Leaky Integrator Model

A neuron's membrane potential can be modeled as a leaky capacitor that gives rise to the differential equation

$$C_m \frac{dV}{dt} = I(t)$$

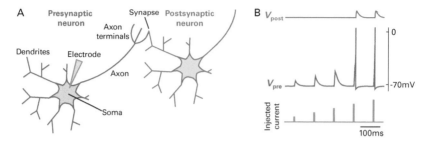

Figure 1.1
A sketch of the dynamics modeled in chapter 1. (A) A diagram of two neurons. An electrode (gray) injects electrical current into the presynaptic neuron (red), which connects to the postsynaptic neuron (blue) at a synapse. (B) The membrane potentials (V_{pre} and V_{post}) of the two neurons in response to current pulses injected into the presynaptic neuron. Sufficiently strong injected current evokes action potentials or "spikes" in the presynaptic neuron's membrane potential, which then evoke responses in the postsynaptic neuron's membrane potential.

where $V(t)$ is the membrane potential, $I(t)$ is current across the membrane, and C_m is the capacitance of the membrane. You may have seen this equation in physics class when studying resistor-capacitor (RC) circuits. In reality, $I(t)$ represents the average current *per-unit area* of the membrane, but we will ignore that detail and simply interpret it as a single current. A current that increases the membrane potential ($I > 0$) is called an *inward* or *depolarizing* current. A current that decreases the membrane potential ($I < 0$) is called an *outward* or *hyperpolarizing* current. Note that when *negative* ions flow *out* of a cell, it is called an *inward current* because the flow of net charge is inward. Similarly, negative ions flowing into a cell is called an *outward current*.

Ions pass through the membrane primarily through two mechanisms: ion pumps and ion channels. Ion pumps use energy to maintain ion concentration differences across the cell membrane. A common example is the sodium-potassium (Na-K) pump, which pumps two K^+ ions in for each three Na^+ pumped out, giving a net-negative current. The Na-K pump is the primary mechanism through which neurons maintain a negative potential.

Ion channels can be thought of as pores that allow ions to pass through membrane. Ions diffuse through on their own due to the concentration gradient and the electrical gradient (unlike pumps where ions are forced through). Different types of channels admit different types of ions to pass through. There are thousands of types of channels in the brain.

For most cortical neurons, the overall effects of many ion channels and pumps when the membrane potential is near rest (V around $-70\,\mathrm{mV}$) can be approximated by a single current called the *leak current*, which is defined by

$$I_L = -g_L(V - E_L). \tag{1.1}$$

The constant $g_L > 0$ is called the *leak conductance* and the constant E_L is the equilibrium or *resting potential* of the neuron. The idea is that different ion pumps and ion channels pull V in different directions with different strengths. The equilibrium potential, E_L, is the value of V at which all of these forces cancel out, so there is no current (because $I_L = 0$ when $V = E_L$). When $V > E_L$, we say that the membrane potential is *depolarized*; when $V < E_L$, we say that it is *hyperpolarized*. Depolarized membrane potentials produce negative currents ($I_L < 0$ when $V > E_L$) that pull the membrane potential back down toward E_L. Similarly, hyperpolarized membrane potentials produce positive currents ($I_L > 0$ when $V < E_L$) that pull the membrane potential back down toward E_L. Hence, the leak current always pulls V toward E_L. The conductance, g_L, measures how strongly V is pulled toward E_L.

In addition to the leak current, we might want to model the current injected by a scientist's electrode (as in figure 1.1a) or some other current source. To this end, we define the total membrane current as

$$I = I_L + I_x$$

where I_x is any external source of current that we want to model. We will sometimes refer to I_x as the neuron's *external input current*. Putting all of this together gives the *leaky integrator model*:

$$C_m \frac{dV}{dt} = -g_L(V - E_L) + I_x(t).$$

Despite its simplicity, this model is a decent approximation of the subthreshold or passive properties of some neurons: that is, the behaviour of $V(t)$ below the spiking threshold and therefore in the absence of spikes. The model can be simplified by setting

$$\tau_m = \frac{C_m}{g_L}$$

and rescaling the input current by taking $I_x \leftarrow I_x/g_L$ to get

> **The leaky integrator model**
>
> $$\tau_m \frac{dV}{dt} = -(V - E_L) + I_x(t). \tag{1.2}$$

The parameter, τ_m, is called the *membrane time constant* and sets the timescale of the membrane potential dynamics. Typical cortical neurons have membrane time constants around 5–20 ms.

Note that $I_x(t)$ is, strictly speaking, not a current in equation (1.2) because it has dimensions of electrical potential (the same as $V(t)$), typically measured in millivolts. This is due to our rescaling by g_L. However, we will still refer to it as a "current" since it is proportional to the actual external current and it still models a current.

> See appendixes A.1 and A.2 for a review of ordinary differential equations (ODEs) for exponential decay.

We next derive and interpret solutions to equation (1.2), first for the case of time-constant input, $I_x(t) = I_0$. When $I_x(t) = I_0$, equation (1.2) is an autonomous, linear differential equation, which has a solution given by

$$V(t) = (V_0 - E_L - I_0)\, e^{-t/\tau_m} + E_L + I_0 \tag{1.3}$$

where $V(0) = V_0$ is the initial condition. Equation (1.3) represents an exponential decay to $E_L + I_0$. The timescale of this decay is set by τ_m. Roughly speaking, τ_m is the amount of

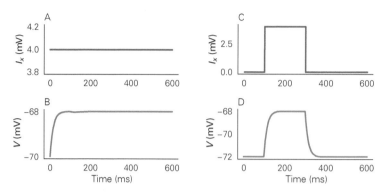

Figure 1.2
Leaky integrator model with time-constant and time-dependent input. (A, B) Time-constant input current, $I_x(t) = I_0 = 4$ mV, and membrane potential, $V(t)$, of a leaky integrator model. (C, D) Same, but with a square-wave time-dependent input, $I_x(t)$ Parameters are $\tau_m = 15$ ms, $E_L = -72$, and $V(0) = -70$. See Leaky Integrator.ipynb for code to produce this figure.

time required for the membrane potential to get a little past halfway from $V(0)$ to $E_L + I_0$ (specifically, it gets a proportion $1 - e^{-1} \approx 0.63$ of the way).

In Python, we can implement the solution in equation (1.3) as

```
V=(V0-EL-I0)*exp(-time/taum)+EL+I0
```

where time is a NumPy array representing discretized time. See figure 1.2a,b and the first code cell of LeakyIntegrator.ipynp for a complete simulation of the leaky integrator with time-constant input.

We next consider the leaky integrator with a time-dependent input, $I_x(t)$. In this case, equation (1.2) is an inhomogeneous, linear differential equation. The solution to equation (1.2) can be written as a convolution of $I_x(t)$ with a kernel.

See appendixes A.3 and A.4 for a review of convolutions and linear ODEs with time-dependent forcing.

Specifically, the solution can be written as

$$V(t) = (V_0 - E_L)\,e^{-t/\tau_m} + E_L + (k * I_x)(t) \tag{1.4}$$

where $V_0 = V(0)$, $*$ denotes convolution, and the kernel is defined by

$$k(s) = \begin{cases} \frac{1}{\tau_m}e^{-s/\tau_m} & s \geq 0 \\ 0 & s < 0 \end{cases}$$

$$= \frac{1}{\tau_m}e^{-s/\tau_m}H(s).$$

Here,

$$H(s) = \begin{cases} 1 & s \geq 0 \\ 0 & s < 0 \end{cases}$$

is the Heaviside step function. We have implicitly assumed that $I_x(t) = 0$ for $t < 0$: that is, the input starts at $t = 0$.

When $V_0 = E_L$ (or $t \gg \tau_m$), the first term in equation (1.4) can be ignored such that

$$V(t) = E_L + (k * I_x)(t). \tag{1.5}$$

In other words, the membrane potential is a filtered version of the input (plus E_L). Since $k(s) = 0$ for $s < 0$, this is a causal filter, meaning that $V(t_0)$ only depends on values of $I_x(t)$ in the past ($t < t_0$). Also, $\int k(t)dt = 1$, so $V(t)$ is a running, weighted average of the recent history of $I_x(t)$. The membrane time constant, τ_m, determines how quickly the neuron responds to changes in $I_x(t)$, equivalently how far in the past it averages $I_x(t)$. Another way of thinking about equation (1.5) is that changes to $I_x(t)$ get integrated into $V(t)$ and then forgotten over a timescale of τ_m.

In Python, we would implement the solution as

```
# Define the convolution kernel
s=np.arange(-5*taum,5*taum,dt)
k=(1/taum)*np.exp(-s/taum)*(s>=0)
# Define V by convolution
V=EL+np.convolve(Ix,k,mode='same')*dt
```

Note that we defined the kernel, $k(s)$, over time interval $[-5\tau_m, 5\tau_m]$ because the time window must be centered at $s = 0$ and because $k(s)$ is close to zero outside this window (i.e., the window contains most of the "mass" of $k(s)$). A complete example of the leaky integrator with time-dependent input is given in figure 1.2c,d and the second code cell of Leaky Integrator.ipynb. Try different time series for $I_x(t)$ to get an intuition for the model.

The leaky integrator does a reasonable job of describing subthreshold (nonspiking) membrane dynamics, but it does not capture action potentials, which occur when the membrane potential exceeds a threshold around -55 mV. Referring back to figure 1.1b, the leaky integrator model captures the membrane potential responses to the first three current pulses, but does not capture the action potentials in response to the last two pulses. In appendix B.1, we discuss the Hodgkin-Huxley (HH) model, which describes how sodium and potassium ion channels produce action potentials. In the next section, we discuss a simpler model that captures many of the salient features of action potential generation.

1.2   The EIF Model

When the membrane potential of a neuron gets close to -55 mV, sodium channels begin opening, causing an influx of sodium, which pushes the membrane potential higher (because sodium is positively charged). This causes more sodium channels to open, creating a positive feedback loop and a rapid upswing in the membrane potential. This rapid upswing is eventually shut down by the closing of sodium channels and the opening of potassium channels that pulls the membrane potential back down toward rest. The HH model from appendix B.1 captures these dynamics in great detail, but it is complicated and computationally expensive to simulate, so we focus on a simplified model here.

The upswing of action potentials and the effects of subthreshold sodium currents can be captured by adding an exponential term to the leaky integrator model. The subsequent "reset" of the membrane potential back toward rest can be captured by a simple rule: Every time $V(t)$ exceeds some threshold, V_{th}, we record a spike and reset $V(t)$ to V_{re}. Putting this together gives the *exponential integrate-and-fire (EIF) model* (Fourcaud–Trocme et al. 2003),

---

The EIF model

$$\tau_m \frac{dV}{dt} = -(V - E_L) + De^{(V-V_T)/D} + I_x(t)$$

$$V(t) > V_{th} \Rightarrow \text{spike at time } t \text{ and } V(t) \leftarrow V_{re}.$$

(1.6)

The second line in equation (1.6) defines the resetting of the membrane potential described here, and it defines the class of *integrate-and-fire* (IF) models. The term

$$\Phi(V) = De^{(V-V_T)/D}$$

models the current induced by the opening of sodium channels that initiates the upswing of an action potential.

The parameter V_T should be chosen near the potential at which action potential initiation begins—that is, where sodium channels begin opening more rapidly ($V_T \approx -55$ mV). Parameters for the EIF must satisfy $V_{re}, V_T < V_{th}$ and $D > 0$. Typically, V_{re} is near E_L. If we want to faithfully model the peak of an action potential, then we should choose $V_{th} \approx 0 - 10$ mV, but choosing smaller values will not greatly affect spike timing.

A simpler model called the leaky integrate-and-fire (LIF) model omits the $\Phi(V)$ term and instead records a spike and resets the membrane potential after V crosses $V_T \approx -55$ mV. The LIF is one of the most widely used models in computational neuroscience. The LIF and other simplified models are discussed in appendix B.2.

A detailed mathematical analysis shows that the membrane potential dynamics of the more detailed Hodgkin-Huxley model during the upswing of an action potential are accurately approximated by the EIF model (see Jolivet, Lewis, and Gerstner (2004) or chapter 5.2 of Gerstner et al. (2014) for details). And careful experiments indicate that the EIF model accurately captures the upswing of an action potential in real neurons (Badel et al. 2008). For these and other reasons, the EIF model is sometimes regarded as an ideal one-dimensional neuron model (Fourcaud–Trocme et al. 2003; Jolivet et al. 2004; Gerstner et al. 2014) (with "one-dimensional" meaning that the model is defined by a one-dimensional ODE).

See appendix A.5 for a review of the forward Euler method.

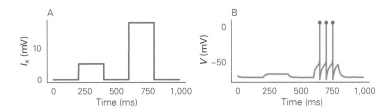

Figure 1.3
Simulation of an EIF neuron model. (A) The input current and (B) the membrane potential of an EIF neuron model. Red circles indicate the times at which the neuron spiked. See `EIF.ipynb` for code to produce this figure.

Unlike the leaky integrator model, we cannot write an equation for the solution to equation (1.6). Instead, we can find an approximate, numerical solution to equation (1.6) using the forward Euler method augmented with a threshold-reset condition:

```
for i in range(len(time)-1):
    # Euler step
    V[i+1]=V[i]+dt*(-(V[i]-EL)+DeltaT*np.exp((V[i]-VT)/DeltaT)
    +Ix[i])/taum
    # Threshold-reset condition
    if V[i+1]>=Vth:
        V[i+1]=Vre
        V[i]=Vth  # This makes plots nicer
        SpikeTimes=np.append(SpikeTimes,time[i+1])
```

See figure 1.3 and `EIF.ipynb` for a complete simulation. The line `V[i]=Vth` sets the membrane potential to V_{th} just before it is reset to V_{re} in the event of a spike. This is added to make the plots of $V(t)$ look nicer, but it has no effect on spike timing or dynamics. Try commenting out this line in `EIF.ipynb` to understand why it helps.

To better understand the dynamics of the EIF model, we begin with an intuitive explanation and then perform a more precise analysis. The intuitive explanation proceeds by considering two regimes of V:

- **The leaky integrator regime.** When $V \ll V_T$ (meaning V is "way less than" V_T), then $\Phi(V) \approx 0$ so the EIF model behaves like a leaky integrator. The membrane potential is pulled toward $E_L + I_x$. The first 600 ms in figure 1.3 illustrates the leaky integrator regime.

- **Spiking regime.** If V is larger than V_T, then $\Phi(V)$ gets larger and produces an inward current that models the opening of sodium channels. For sufficiently large V, this inward current creates a positive feedback loop: Increasing V causes $\Phi(V)$ to increase, which increases V further, causing $\Phi(V)$ to increase further and so on. This feedback loop models the upswing of the membrane potential at the initiation of an action potential. The EIF in figure 1.3 is pushed into the spiking regime by the larger input pulse beginning at 600 ms.

There is an intermediate regime in which the inward current, $\Phi(V)$, is appreciably larger than zero but is not strong enough to produce a positive feedback loop and initiate an action potential by itself. Instead, $\Phi(V)$ reduces the amount of positive input, $I_x > 0$, required to initiate an action potential. In this sense, $\Phi(V)$ implements a kind of *soft threshold* for the EIF model near V_T: When $V > V_T$, an action potential is likely to occur, but it is not guaranteed.

Parameter D determines how soft the soft threshold is: that is, how gradually the inward current increases as V increases toward V_T and beyond. When D is small, $\Phi(V) \approx 0$ whenever V is even just a little bit below V_T, but $\Phi(V)$ is very large when V is just a little bit above V_T. This leads to a sharper threshold near V_T and a faster action potential upswing. When D is larger, $\Phi(V)$ increases more gradually when V is near V_T, leading to a softer threshold near V_T and a slower action potential upswing. Typically, we should choose $D \approx 1 - 5$ mV.

> See appendix A.6 for a review of phase lines, fixed points, and stability for one-dimensional ODEs.

To understand the dynamics of the EIF more precisely, we can consider the constant input case, $I_x(t) = I_0$, and perform a phase line analysis. The subthreshold dynamics of the EIF with time-constant input obey

$$\frac{dV}{dt} = f(V)$$

where

$$f(V) = \frac{-(V - E_L) + De^{(V - V_T)/D} + I_0}{\tau_m}.$$

The phase line for $I_0 = 5$ mV is plotted in figure 1.4a. There is a stable fixed point near $E_L + I_0$ and an unstable fixed point just above V_T. Therefore, so long as $V(0) < V_T$, no spike will occur because $V(t)$ will converge to the fixed point near $E_L + I_0$, similar to the leaky integrator model. Note also that $f(V)$ is approximately linear near $E_L + I_0$, because $\Phi(V)$ is small there, so the EIF behaves like the leaky integrator in this regime.

Increasing I_0 is equivalent to shifting up the blue curve, $f(V)$, in figure 1.4. When we increase the input to $I_0 = 20$ mV (figure 1.4b), both fixed points disappear because $f(V) > 0$ for all V. Regardless of the initial condition, the membrane potential will increase to V_{th} then

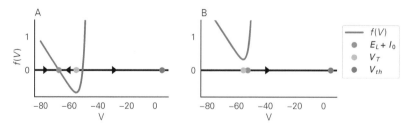

Figure 1.4
Phase line for the EIF model. (A) Subthreshold regime with $I_0 = 5$ mV. (B) Superthreshold regime with $I_0 = 20$ mV. Parameters are $\tau_m = 15$ ms, $E_L = -72$, $V_{th} = 5$, $V_{re} = -75$, $V_T = -55$, and $D = 2$. See EIFphaseLine.ipynb for code to produce this figure.

reset to V_{re} and this process will repeat. Note that $f(V)$ grows exponentially when $V > V_T$, which means that $V'(t) = f(V)$ increases very rapidly whenever $V > V_T$. This explains the fast upswing in the action potentials in figure 1.3b.

The existence of a stable and unstable fixed point when $I_0 = 5$ mV and their disappearance when $I_0 = 20$ mV in figure 1.4 is a signature of a saddle-node bifurcation. Define I_{th} as the value of I_0 at which the bifurcation occurs (that value at which the two fixed points collide). This value allows us to precisely characterize the two regimes for an EIF with time-constant input.

1. **Subthreshold regime:** If $I_0 \leq I_{th}$, the membrane potential decays exponentially toward a fixed point and the neuron never spikes.

2. **Superthreshold regime:** If $I_0 > I_{th}$, the membrane potential eventually reaches V_{th}, the neuron spikes, V is reset to V_{re}, and this process repeats indefinitely.

For these reasons, I_{th} is the *threshold input*: that is, the input strength necessary to drive the EIF to spike. The subthreshold and superthreshold regimes are demonstrated by the response to the two inputs given in figure 1.3. Interestingly, despite the fact that the fixed point in the subthreshold regime cannot be derived as a closed-form expression, the threshold input can be written in closed form.

Exercise 1.2.1. The threshold input, I_{th}, of the EIF with time-constant input is defined as the value at which the neuron eventually spikes if $I_0 > I_{th}$, but never spikes if $I_0 \leq I_{th}$. Derive a closed-form equation for I_{th} for the EIF model. To verify your solution, plot the phase line and numerical simulations of the EIF model for $I_0 = I_{th} - 1$ mV and $I_0 = I_{th} + 1$ mV.

Hint: The minimum of $f(V)$ can be found by setting $f'(V) = 0$, solving for V, then plugging this value of V back into $f(V)$. From figure 1.4, we can see that the EIF spikes whenever the minimum of $f(V)$ is larger than zero. In your derivation, you can make the reasonable assumption that E_L is sufficiently smaller than V_T.

Exercise 1.2.2. The EIF model with a superthreshold, time-constant input, $I_x(t) = I_0 > I_{th}$, spikes periodically. The frequency of spikes is called the "firing rate," which is sometimes denoted as r. The rate can be estimated by counting the number of spikes and dividing by the duration, T, of the simulation:

$$r = \frac{\text{\# of spikes}}{\text{duration of simulation}}$$

Make a plot of r as a function of I_0. This is called an *f-I curve* (for "frequency-input"). Use values of I_0 starting from $I_0 < I_{th}$ and ending near where $r = 50$ Hz.

Hint: Create a vector, `I0s`, of values for `I0`. Write a for loop that simulates the EIF for each value of `I0` in `I0s`. The simulation inside this loop should look like the code in `EIF.ipynb`. After each simulation, compute r and store it in a separate vector, `rs`. Then plot `rs` versus `I0s`. Note that time is measured in ms, so $r = 50$ Hz corresponds to a value of $r = 0.05$ (implicitly measured in kilohertz).

1.3 Modeling Synapses

We have so far considered very simple forms of external input, $I_x(t)$, modeling current injected by an electrode. We will next model currents that come from other neurons. A *synapse* is a connection between two neurons, the *presynaptic* neuron and *postsynaptic* neuron (see figure 1.1 and the surrounding discussion). There are two fundamental types of synapses: ionotropic and metabotropic. Ionotropic synapses are faster and more direct than metabotropic. We will focus on ionotropic synapses in this book and will not consider metabotropic synapses.

When the presynaptic neuron spikes, neurotransmitter molecules are released and diffuse across the synapse to receptors on the postsynaptic neuron's membrane. These neurotransmitters open ion channels, which evokes a current across the postsynaptic neuron's membrane, called a *postsynaptic current (PSC)*. The PSC evoked by each presynaptic spike causes a transient response in the membrane potential of the postsynaptic neuron, called a *postsynaptic potential (PSP)*. In figure 1.1b, the two bumps in the blue curve are PSPs evoked by the two spikes in the red curve.

Our goal in this chapter is to develop a model of the synaptic dynamics sketched in figure 1.1b. We will consider a synapse model in which currents are generated directly from presynaptic spike times. This is called a *current-based synapse model*. A more biologically detailed model generates a synaptic *conductance* from spike times and the synaptic current is generated from this conductance. These conductance-based synapse models are discussed in appendix B.3.

There are two polarities of ionotropic synapses: *excitatory* and *inhibitory*. Excitatory synapses evoke positive (inward) synaptic currents, pushing the membrane potential closer to threshold, thereby "exciting" the neuron. Inhibitory synapses evoke negative (outward) synaptic currents, pushing the membrane potential away from threshold, thereby "inhibiting" action potentials.

Dale's Law says that a presynaptic neuron connects to all its postsynaptic targets with the same type of synapse. For our purposes, this will be taken to mean that all postsynaptic targets of a single neuron are either excitatory or inhibitory. Laws in neuroscience are almost never completely universal, but Dale's law has very few exceptions. When considering cortical neurons of adult mammals under healthy conditions, one can safely apply Dale's law.

Because of Dale's law, we can classify *neurons* as excitatory neurons or inhibitory neurons, instead of just classifying *synapses* as excitatory or inhibitory. An excitatory neuron is a neuron that connects to all its postsynaptic targets with excitatory synapses, and similarly for inhibitory neurons. In cortex, about 80 percent of neurons are excitatory neurons and 20 percent are inhibitory. Most excitatory neurons in the cortex are pyramidal neurons, named for the pyramid-like appearance of their soma. Most inhibitory neurons in the cortex are classified as cortical interneurons, where the "interneuron" label reflects the fact that they only project locally, to nearby neurons in the same cortical area or layer. In the cortex, long-range projections to other cortical areas are almost exclusively made by excitatory (pyramidal) neurons.

We begin by modeling a leaky integrator with one excitatory and one inhibitory synapse:

$$\tau_m \frac{dV}{dt} = -(V - E_L) + I_e(t) + I_i(t)$$

$$I_e(t) = J_e \sum_j \alpha_e(t - s_j^e)$$

$$I_i(t) = J_i \sum_j \alpha_i(t - s_j^i)$$

(1.7)

The functions $I_e(t)$ and $I_i(t)$ are excitatory and inhibitory *synaptic currents*. The $\alpha_e(t)$ and $\alpha_i(t)$ are called *PSC waveforms*. The sign and magnitude of the synaptic currents are determined by the *synaptic weights*, $J_e > 0$ and $J_i < 0$. Equation (1.7) can be interpreted as adding a copy of the PSC waveform, $\alpha_a(t)$, scaled by the weight, J_a, at each presynaptic spike time, s_j^a, for $a = e, i$. Figure 1.5b shows the synaptic current generated by presynaptic spikes in figure 1.5a.

Which function should we use for the PSC waveforms? We must take $\alpha_a(t) = 0$ for $t < 0$ because the postsynaptic response cannot precede the presynaptic spike. We can also assume that $\alpha_a(t) \geq 0$ and $\int \alpha_a(t)dt = 1$ since we can absorb the sign and integral of $\alpha_a(t)$ into J_a. A widely used and simple model of a PSC waveform is an exponential decay (as in figure 1.5b,d),

$$\alpha_a(t) = \frac{1}{\tau_a} e^{-t/\tau_a} H(t)$$

(1.8)

for $a = e, i$ where

$$H(t) = \begin{cases} 1 & t \geq 0 \\ 0 & t < 0 \end{cases}$$

is the Heaviside step function and τ_a is the *synaptic time constant* that controls how quickly the PSC decays. The $1/\tau_a$ factor in the definition of $\alpha_a(t)$ ensures that $\int \alpha_a(t)dt = 1$. This form of $\alpha_a(t)$ is sometimes called an *exponential synapse model*. Synaptic timescales of many ionotropic synapses in the cortex are around $\tau_a \approx 5 - 10$ ms.

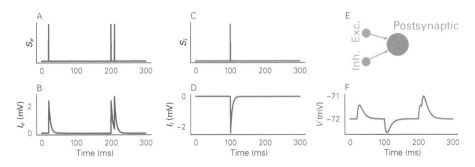

Figure 1.5
A leaky integrator model driven by two synapses. (A) Excitatory spike density. Each vertical bar represents a spike in the excitatory presynaptic neuron, modeled as a Dirac delta function. (B) Excitatory synaptic current generated by the spikes in A using an exponential synapse model. (C) Inhibitory spike density. (D) Inhibitory synaptic current. (E) Schematic of an excitatory and inhibitory neuron connected to a postsynaptic neuron. (F) Membrane potential of the postsynaptic neuron modeled as a leaky integrator. See Synapses.ipynb for code to produce this figure.

Figure 1.5f shows the membrane potential generated by this model. Each isolated presynaptic spike evokes a PSP. PSPs from excitatory presynaptic spikes are called *excitatory postsynaptic potentials (EPSPs)* and those from inhibitory presynaptic spikes are called *inhibitory postsynaptic potentials (IPSPs)*. The *PSP amplitude* is the height of a PSP (the distance from the peak of the PSP to the resting membrane potential). PSP amplitudes in the cortex are often between 0 and 1 mV. PSP amplitudes are proportional to the synaptic weight, J_e or J_i, but they also depend on the synaptic timescales, τ_e or τ_i, and membrane time constant, τ_m. When building simulations, it's often easiest to choose the timescales and time constants first, and then choose the synaptic weights by trial-and-error to get the PSP amplitude you want.

See appendix A.7 for a review of Dirac delta functions.

Before describing how to *simulate* the model in equation (1.7) to generate figure 1.5, we first describe how to reformulate the model to make it easier to simulate. First, we write the presynaptic spike train as a sum of Dirac delta functions:

$$S_e(t) = \sum_j \delta(t - s_j^e)$$

$$S_i(t) = \sum_j \delta(t - s_j^i). \tag{1.9}$$

This representation of a spike train is called a *spike density*. Specifically, the spike density representation of a spike train is a time series with a Dirac delta function at each spike time. Spike density representations of spike trains can simplify mathematics and coding in many situations in computational neuroscience. Figures 1.5a,c show spike density representations of the presynaptic spike trains in which vertical bars are used to represent Dirac delta functions. Python code for turning a list of spike times into a spike density is given by

```
Se=np.zeros_like(time)
Se[(ExcSpikeTimes/dt).astype(int)]=1/dt
```

where `ExcSpikeTimes` is a vector of spike times. This code sets indices corresponding to spike times to the value `1/dt`, representing a Dirac delta function in discrete time.

Using the spike density representation of the presynaptic spike train allows us to write the synaptic currents in ways that are easier to compute. One approach is to write them in terms of convolutions:

$$I_a(t) = J_a(\alpha_a * S_a)(t). \tag{1.10}$$

When $S_a(t)$ are spike densities, equation (1.10) gives the same synaptic currents as equation (1.7). This approach works for any choice of $\alpha_a(t)$.

For the exponential synapse models, there is a simpler approach. The exponential decay of $I_e(t)$ and $I_i(t)$ that occurs between spikes can be represented by a linear ODE. Therefore, the synapse model can be defined by a linear ODE, with the added condition that we increment the conductance at each presynaptic spike. The spike density representation again makes the model easy to formulate. Specifically, the model can be written as

> **Leaky integrator with excitatory and inhibitory synapses**
>
> $$\tau_m \frac{dV}{dt} = -(V - E_L) + I_e(t) + I_i(t)$$
>
> $$\tau_e \frac{dI_e}{dt} = -I_e + J_e S_e(t) \tag{1.11}$$
>
> $$\tau_i \frac{dI_i}{dt} = -I_i + J_i S_i(t)$$

Think about why equation (1.11) is equivalent to equation (1.7) whenever we use an exponential PSC kernel and take initial conditions $I_e(0) = I_i(0) = 0$. With equation (1.11), we can simulate a leaky integrator with synaptic inputs using a simple forward Euler solver. The full code to generate figure 1.5 using this approach is given in `Synapses.ipynb`.

In this section, we only considered a neuron with a single excitatory and inhibitory synapse. Neurons in the cortex receive hundreds or thousands of synaptic inputs. In the following chapters, we will model neurons receiving a large number of synaptic inputs. But first we need to improve how we model the *timing* of presynaptic spikes, which is the topic of chapter 2.

Exercise 1.3.1. Some studies use *instantaneous* or delta synapses, which cause an instant jump in $V(t)$. These can be obtained by taking $\alpha_a(t) = \delta(t)$ or, more simply, by taking $I_a(t) = J_a S_a(t)$ so equation (1.11) becomes

$$\tau_m \frac{dV}{dt} = -(V - E_L) + J_e S_e(t) + J_i S_i(t)$$

and we do not need to compute I_e or I_i at all. Reproduce figure 1.5f, but replace the exponential PSCs with delta synapses. Visually compare the membrane potential traces in the two cases.

2

Measuring and Modeling Neural Variability

2.1 Spike Train Variability, Firing Rates, and Tuning

Figure 2.1a shows a recording of a membrane potential from a real neuron in the brain of a rat[1] (Okun, Naim, and Lampl 2010). Some features match what we saw in chapter 1: The membrane potential hovers near -70 mV, with the exception of brief action potentials to near 0 mV. However, there are other features that are noticeably different. Most notably, the membrane potential fluctuates in a seemingly random fashion between spikes, and the spike times are irregular. These types of irregularity are ubiquitous in recordings made from living animals and are broadly known as *neural variability*. The causes and effects of neural variability are a topic of active research and debate.

To model and quantify neural variability, we will begin by studying the irregularity of spike times, which is often called *spike timing variability*. To begin studying neural variability, we first define the *spike count* over a time interval:

Spike count definition

$$N(a,b) = \# \text{ of spikes in } [a,b].$$

Given a single recording, we can compute the spike count over the entire recording interval, such as $N(0, 5000) = 11$ for the recording in figure 2.1a. We can also compute the spike counts over consecutive intervals of a given length. The following code computes firing rates over consecutive intervals of length 500 ms:

```
dtRate=500
SpikeCountTime=np.arange(0,5001,dtRate)
SpikeCounts=np.histogram(SpikeTimes,SpikeCountTime)[0]
```

---

1. Thanks to Micheal Okun and Ilan Lampl for sharing the data shown in figure 2.1. More information about the data and its collection can be found in Okun, Naim, and Lampl (2010).

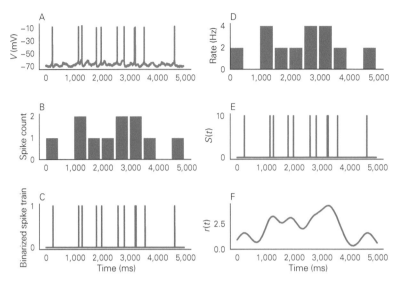

Figure 2.1
Membrane potential, spike counts, and time-dependent firing rate of a real neuron. (A) The membrane potential recorded from rat cortical neuron. The potential was shifted downward to correct for possible recording artifacts. (B) Spike counts over consecutive 500-ms intervals. (C) Binarized spike count using time bins of width $dt = 0.1$ms. (D) Time-dependent firing rate computed over consecutive 500-ms intervals. (E) Numerical spike density using $dt = 0.1$ms. (F) A smooth, time-dependent firing rate estimate computed by convolving with a Gaussian kernel. See `OneRealSpikeTrain.ipynb` for the code to produce this figure.

The call to the `histogram` function counts the number of elements in `SpikeTimes` between every pair of consecutive elements in `SpikeCountTime`. The results are shown in figure 2.1b.

As discussed in chapter 1, a neuron will not typically spike twice within a 2-ms time window. Therefore, if we discretize time into very small intervals, $dt \leq 2$ ms, then each bin should contain either 1 spike or zero spikes. This is called a *binarized spike train* because it is a binary time series (0s and 1s). See figure 2.1c for an example of a binarized spike train with bin width $dt = 0.1$ ms.

We're often interested in the frequency of spikes—that is, the number of spikes per unit time—which is called a neuron's *firing rate* or just *rate*:

Firing rate definition

$$r = \text{firing rate} = \frac{\text{\# of spikes}}{\text{length of time window}} = \frac{N(a,b)}{b-a}.$$

Firing rates have the dimension "number of spikes per unit time." If we think of "number of spikes" as dimensionless, then r is a frequency. If time is measured in millisecond, then r has units of "spike per ms" or kilohertz. But we usually report firing rates in units of spikes per second or hertz. Since we usually keep track of time in milliseconds in our code, this often requires us to rescale firing rates to hertz by multiplying by 1,000 (because 1 kHz = 1,000 Hz). Neurons in the cortex usually spike at around 1–50 Hz.

Each measure of the spike count discussed here has an analog in firing rates. As with spike counts, we can measure the firing rate over an entire recording or simulation, which can be called the *time-averaged firing rate*. For example, the time-averaged rate in the recording from figure 2.1a is 2.2 Hz. We can also measure the firing rate over sequential time intervals, which is called a *time-dependent firing rate*, such as the time-dependent rate plotted in figure 2.1d. Note that the firing rates in figure 2.1d are just the spike counts from figure 2.1b scaled by a factor of 2 since the time intervals are $b - a = 0.5$ s long. If we measure firing rates over small time intervals, $dt < 2$ ms, then each bin in the time-dependent rate should contain either zero or $1/dt$. This is just a discrete-time representation of the spike density:

$$S(t) = \sum_j \delta(t - s_j)$$

where s_j is the jth spike time. Figure 2.1e shows the time-dependent firing rate (or discretized spike density) with a time bin size of $dt = 0.1$ ms.

The firing rate estimate in figure 2.1d is discontinuous and jagged because we counted spikes over nonoverlapping intervals. We often want an estimate of the firing rate that is continuous in time. One option is to use overlapping time intervals. A more parsimonious example is to convolve the spike density with a smoothing kernel:

$$r(t) = (k * S)(t)$$

where $k(t)$ is a kernel satisfying $\int k(t)dt = 1$. The resulting time series, $r(t)$, is also called a "time-dependent firing rate," but it is smoother than the one defined by counting spikes over disjoint time intervals. Using a Gaussian-shaped kernel is common and can be accomplished in code as follows:

```
sigma=250
s=np.arange(-3*sigma,3*sigma,dt)
k=np.exp(-(s**2)/(2*sigma**2))
k=k/(sum(k)*dt)
SmoothedRate=np.convolve(k,S,'same')*dt
```

Figure 2.1f shows the smoothed rate obtained by applying this method to the spike times.

If you think that all this analysis seems excessive for the handful of spikes in figure 2.1, you're not wrong. More commonly, a neuron will be recorded for a longer duration or it will be recorded across repetitions of the same experiment (*e.g.*, several presentations of the same stimulus). Each repetition is sometimes called a *trial*.

The data file `SpikeTimes1Neuron1Theta.npz` contains spike times estimated recorded from a neuron in a monkey's visual cortex over the course of 200 trials[2] (Smith and Kohn 2008). During each 1-s trial, the monkey watched the same "drifting grating" movie, in which angled bars drifted across part of the monkey's visual field (see figure 2.2a for a still image of a drifting grating). The file has a variable `theta` representing the angle of

---

2. Thanks to Adam Kohn and Matthew Smith for sharing this data, which was recorded from an array of extracellular electrodes. More information about the data and its collection can be found in Smith and Kohn (2008).

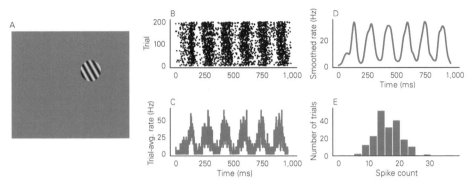

Figure 2.2
Analysis of spike trains recorded from a real neuron across 200 **trials.** (A) Example of a drifting grating visual stimulus. The white/black lines drift slowly across the circle. (B) Raster plot; each dot is a spike at the indicated time and trial. (C) Time-dependent, trial-averaged firing rate computed by counting the number of spikes across all neurons in each time bin. (D) Smoothed, trial-averaged firing rate. (E) Histogram of spike counts across trials. The code to produce this figure can be found in `MultiTrialSpikeTrains.ipynb`.

the stimulus for this recording, which is 120 degrees. Neurons were recorded using an array of extracellular electrodes. A method called "spike sorting" was used to determine which detected spikes were emitted by the neuron in question. While spike sorting is not perfect, we will work under the assumption that spikes were accurately sorted.

The spike times of every spike across all trials is stored in `SpikeTimes`, and the corresponding trial numbers are stored in `TrialNumbers`. For example, `Spike Times[2]==6.53` and `TrialNumbers[2]==157`, indicating that there was a spike 6.53 ms after the start of trial 157. Figure 2.2b shows a *raster plot* of the spike times. In a raster plot, each dot represents a spike at the corresponding time and trial. A raster plot can be generated as follows:

```
plt.plot(SpikeTimes,TrialNumbers,'.')
```

The list of spike times and trial numbers is a memory-efficient way to store multiple spike trains, but that can make it difficult to perform operations on the data. Another option is to use an array of spike densities. Mathematically, we can think of this as a time-dependent vector, $S(t)$, of spike densities where the kth entry of the vector represents the spike density of the kth trial:

$$S_k(t) = \sum_j \delta(t - s_j^k).$$

In NumPy, S can be represented with an array in which `S[k,:]` is the kth spike density. This array can be created as follows:

```
S=np.zeros((NumTrials,len(time)))
S[TrialNumbers,(SpikeTimes/dt).astype(int)]=1/dt
```

Once this array is created, it becomes very easy to compute firing rates by taking averages of S across time and/or trials. The *time-averaged, trial-averaged firing rate* is a single number computed by

```
TrialAvgTimeAvgRate=np.mean(S)
```

For the data shown in figure 2.2, we get 16.2 Hz. We can alternatively compute the *time-dependent, trial-averaged firing rates* by averaging over trials only:

```
TrialAvgRates=np.mean(S,axis=0)
```

The result is plotted in figure 2.2c. The oscillations apparent in figure 2.2c are caused by the periodic movement of the drifting grating stimulus as it drifts. While the oscillatory trend in figure 2.2c is visible, it appears choppy and noisy. A smoothed firing rate can be obtained by convolving with a Gaussian kernel. You might be tempted to convolve each S[k,:] with a kernel and then average the results, but it is the same (and much simpler) to convolve TrialAvgRates with the same kernel:

```
sigma=10
s=np.arange(-3*sigma,3*sigma,dt)
k=np.exp(-(s**2)/(2*sigma**2))
k=k/(sum(k)*dt)
SmoothedRate=np.convolve(k,TrialAvgRates,'same')*dt
```

The result is plotted in figure 2.2d, and the oscillations are more clearly visible.

So far, we have averaged the data across trials, but figure 2.2b shows clear variability across trials as well. We can compute the spike count on each trial by using a Riemann sum to integrate S across time:

```
SpikeCounts=np.sum(S,axis=1)*dt
```

After running this line of code, SpikeCounts is a vector for which each entry is the spike count on a corresponding trial. The spike count data can be visualized by plotting a histogram:

```
plt.hist(SpikeCounts)
```

The result is plotted in figure 2.2e. Even though the average spike count is 16.2, there is substantial variability across trials. In other words, even though the animal is viewing the same drifting grating stimulus on every trial, the neuron emits a different number of spikes on each trial. This is known as *trial-to-trial variability*, or simply *trial variability*. Trial variability of spike counts is often measured using the *Fano factor*, defined as the ratio between the variance and mean of spike counts across trials:

$$FF = \frac{\text{variance of spike counts}}{\text{mean spike count}} = \frac{\text{var}(N(a,b))}{\text{mean}(N(a,b))}.$$

The Fano factor can be computed over any interval, $[a, b]$, but we often just use the entire recording, $[0, T]$. If a neuron spikes exactly the same number of times on every trial, then $\text{var}(N) = 0$ so $FF = 0$. More generally, a larger Fano factor implies greater trial-to-trial variability.

Neurons encode information about stimuli and behaviour in their spike trains. For example, neurons in the visual cortex change their firing rates in response to changes in a visual stimulus and, similarly, neurons in the motor cortex change their firing rates during motor behaviour. It is widely believed that the encoding of information in neurons' spike trains forms the basis of perception, cognition, and behaviour, but the details of how this happens are not understood.

A classical example of neural coding is *orientation tuning* in the primary visual cortex (V1). In the 1950s and 1960s, David Hubel and Torsten Wiesel discovered that some neurons in cat V1 increased their firing rates when a bar of light passed through a small region of the cat's visual field. Different neurons respond to bars of light in different regions of the visual field. A neuron's *receptive field* is the region to which it responds. Many neurons' firing rates depend on the angle or "orientation" of the bar of light. The orientation that evokes the highest firing rate is called the neuron's *preferred orientation*. Hence, firing rates of neurons in V1 encode the location and orientation of bars of light. The encoding of orientations in V1 is still widely studied today. Hubel and Wiesel won a Nobel Prize in Physiology or Medicine in 1981 for their work.

The spike trains shown in figure 2.2 were recorded in response to a drifting grating at one particular orientation ($\theta = 120$ degrees), but the neuron was actually recorded across twelve different orientations $(0, 30, 60, \ldots, 330$ degrees$)$ with several trials for each orientation. The file `SpikeCounts1Neuron12Thetas.npz` contains an array with the spike counts for all 100 trials on each of the twelve orientations. Trial-averaged firing rates for each orientation can be computed and plotted by

```
Rates=SpikeCounts/T
TrialAvgRates=np.mean(Rates,axis=0)
plt.plot(AllOrientations,1000*TrialAvgRates)
```

The result (figure 2.3a) is called the neuron's *tuning curve* because it depicts how the neuron is "tuned" to each orientation; that is, how the neuron's trial-averaged firing rate is affected by each orientation. Why do there appear to be two preferred orientations? Think about the properties of the drifting grating stimulus to answer this question.

Figure 2.3a shows only the trial-averaged response of the neuron, but we know from the previous discussion that there is trial-to-trial variability. Indeed, figure 2.3b shows

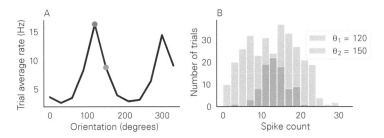

Figure 2.3
Orientation tuning curve and spike count histograms from one neuron. (A) Tuning curve (trial-averaged rate as a function of orientation) for a neuron recorded in monkey visual cortex. Blue and pink dots mark the orientations $\theta_1 = 120$ degrees and $\theta_2 = 150$ degrees. (B) Histogram of spike counts across trials for θ_1 and θ_2. The Code to produce this plot can be found in `OrientationTuningCurve.ipynb`.

histograms of spike counts for two orientations. Let's suppose you tried to infer the orientation of the stimulus by looking at the neuron's spike count on a single trial. You would not be able to correctly infer the orientation on every trial (why?). However, there are millions of neurons in the animal's visual cortex, each with its own tuning curve. If you could observe the spike counts of all of them (and you knew their tuning curves), you could infer the orientation more accurately. The science of understanding how stimuli are encoded in neural activity, and how they can be decoded, is called *neural coding*. Appendix B.4 discusses neural coding in more depth. The last exercise in section 4.2 of chapter 4 uses an artificial neural network (ANN) to decode spike counts from a population of neurons.

Exercise 2.1.1. Compute the Fano factor for the spike trains in figure 2.2b.

2.2 Modeling Spike Train Variability with Poisson Processes

In the previous section, we looked at how to *measure* firing rates and spike timing variability of real, recorded spike trains. In this section, we talk about how to *model* spike timing variability, which will allow us to generate synthetic spike trains and model the effects of spike timing variability in neural circuits.

To account for the variability of real spike trains, we can model them as a stochastic process. A *stochastic process* is similar to a random variable, but it is a random function of time. Specifically, we will model a spike train as a special type of stochastic process called a *point process*, which is a stochastic process representing discrete events in time. Here, those events are spikes. Two common ways to represent point processes are the counting process,

$$n(t) = \text{ \# of spikes in } [0, t] = N(0, t)$$

and the spike density, $S(t)$. Modeling spike trains as stochastic processes allows us to make precise definitions of probabilities, statistics, and other properties. For example, the *instantaneous firing rate* is defined as follows:

Instantaneous firing rate definition
$$r(t) = \lim_{\delta \to 0} \frac{E[N(t, t+\delta)]}{\delta} = \frac{d}{dt} E[n(t)] = E[S(t)] \qquad (2.1)$$

where $E[\cdot]$ denotes expectation. The last equality is difficult to make precise because $S(t)$ is an unusual type of process (being composed of Dirac delta functions), but it will be useful later. Intuitively, $\frac{d}{dt}n(t) = S(t)$ because $n(t) = \int_0^t S(\tau)d\tau$, so $\frac{d}{dt}E[n(t)] = E[n'(t)] = E[S(t)]$. This intuition will have to suffice for our purposes. The spike count, $N(a,b)$, is a random number, and

$$E[N(a,b)] = \int_a^b r(t)dt.$$

A stochastic process is said to be *stationary* if its statistics do not change across time; that is, the statistics of $x(t)$ are the same as those of $y(t) = x(t + t_0)$. For point processes, stationarity can be phrased as follows:

Definition. A point process is stationary if $N(a,b)$ has the same distribution as $N(a + t_0, b + t_0)$ for any t_0 and any $a < b$.

In other words, the distribution of $N(a,b)$ depends only on $b - a$. Therefore, for stationary point processes, we can often just talk about $n(t)$ instead of $N(a,b)$ since they have the same statistics when $t = b - a$. This property gives the following useful theorem:

Theorem. A stationary point process has a constant rate, $r(t) = r$, and therefore $E[n(t)] = rt$.

Poisson processes provide a canonical statistical model of noisy spike trains. The simplest and most common version of a Poisson process is the *homogeneous Poisson process*, sometimes called the *stationary Poisson process*. We will use the term *Poisson process* to mean "homogeneous Poisson process." Inhomogeneous Poisson processes are discussed later in this chapter. A Poisson process is defined as follows:

Definition. A (homogeneous) Poisson process is any stationary point process having the *memoryless property*: $N(t_1, t_2)$ is independent of $N(t_3, t_4)$ whenever $[t_1, t_2]$ is disjoint from $[t_3, t_4]$.

The basic idea is that a spike is equally likely to occur in a Poisson process at any point of time, regardless of how many spikes occur over any interval of time before or after it. This is, of course, not exactly true of real spike trains, but it serves as a simplifying approximation. The Poisson process gets its name from the following theorem:

Theorem. The spike counts of a Poisson process obey a Poisson distribution:

$$\Pr(n(t) = n) = \frac{(rt)^n}{n!}e^{-rt}.$$

A Poisson distribution has the same variance and mean, $\mathrm{var}(n(t)) = E[n(t)] = rt$, which implies that Poisson processes have a Fano factor of 1 over any time window:

$$FF_{Poisson} = \frac{\mathrm{var}(N(a,b))}{E[N(a,b)]} = 1.$$

for any $a < b$. Of course, a sample Fano factor computed from sample Poisson processes will not be exactly equal to 1, but it should converge to 1 as the number of trials goes to ∞, by the law of large numbers. A Fano factor of 1 is interpreted as a baseline amount of trial-to-trial variability. The phrases *subPoisson* and *superPoisson* are sometimes used to refer to spike trains with $FF < 1$ and $FF > 1$, respectively.

There are several algorithms for generating realizations of Poisson processes, and we will consider two of them here. First, an algorithm that generates spike times directly:

Poisson Process Algorithm 1.   To generate spike times of a Poisson process in the interval $[0, T]$, first generate the spike count from a Poisson distribution with mean rT, and then distribute the spike times uniformly in the interval.

In Python, this is implemented by the following code:

```
N=np.random.poisson(r*T)
SpikeTimes=np.sort(np.random.rand(N)*T)
```

The next algorithm generates a spike density, $S(t)$, instead of spike times.

Poisson Process Algorithm 2.   To generate a spike density representation of a Poisson process over the interval $[0, T]$, first make sure that $r * dt$ is small, and then set each bin of $S(t)$ to $1/dt$ independently with probability $r * dt$ and set all other bins to zero.

For a binarized spike train, you would just set the bins to 1 instead of $1/dt$. In Python, this algorithm can be implemented in a single line:

```
S=np.random.binomial(1, r*dt, len(time))/dt
```

Figure 2.4a shows a spike density of a Poisson process. Algorithm 2 assumes that $r * dt$ is small enough so that you can safely ignore the small probability of two spikes being in the same bin.

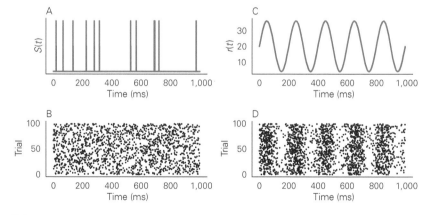

Figure 2.4
Poisson processes. (A) Raster plot of a single Poisson process with $r = 15$ Hz. Each vertical bar is a spike. (B) Raster plot of 100 i.i.d. realizations of a Poisson processes with $r = 15$ Hz. Each dot is a spike. (C, D) Firing rate and raster plot of 100 i.i.d. realizations of an inhomogeneous Poisson process. See `PoissonProcesses.ipynb` for the code to produce these plots.

We can easily switch between spike-time and time-series representations of spike trains as follows:

```
# From spike density to spike times
SpikeTimes=np.nonzero(S)[0]*dt
# From spike times to spike density
S=zeros_like(time)
S[(SpikeTimes/dt).astype(int)]=1/dt
```

To generate multiple i.i.d. trials of a Poisson process (*e.g.*, to model multiple trials), we can just change the size of the generated spike density:

```
S=np.random.binomial(1, r*dt, (NumTrials,len(time)))/dt
```

Figure 2.4b shows a raster plot of 100 trials of a Poisson process with $r = 15$ Hz.

Homogeneous Poisson processes provide a rough approximation to cortical spike train statistics. For example, Fano factors of cortical neurons are often close to 1 (Softky and Koch 1993; Shadlen and Newsome 1994, 1998; Churchland et al. 2010). However, stationary Poisson processes are not an accurate model of recorded spike trains when firing rates change during the recorded time interval, such as in response to time-varying stimuli. For example, in figure 2.2, the firing rate of the neuron oscillates due to the periodic nature of the drifting grating stimulus. In these cases, we should model the spike train as a nonstationary point process. A standard model for nonstationary spike trains is the *inhomogeneous Poisson process*, which is defined as follows:

Definition. Given a nonnegative function, $r(t)$, an inhomogeneous Poisson process with rate $r(t)$ is a point process that has the memoryless property and satisfies

$$E[N(a,b)] = \int_a^b r(t)dt$$

for any $a \leq b$.

Just like the homogeneous Poisson process, the inhomogeneous Poisson process has a Fano factor equal to 1 over any time interval. To generate an inhomogeneous Poisson process, we can generalize Poisson Process Algorithm 2 to a time-dependent rate:

Inhomogeneous Poisson Process Algorithm. First, choose a dt that is small enough so that $r(t) * dt$ is very small for all t. Then set the value of each bin to $1/dt$ with probability $r(t) * dt$, and other bins to zero.

In Python, we can generate spike density representations of inhomogeneous Poisson processes in exactly the same way that we do for homogeneous processes, except that r is a time series (it has the same size as `time`) instead of a scalar. For example, the code here generates an inhomogeneous Poisson process with a sinusoidal rate:

```
r=(20+16*np.sin(2*np.pi*time/200))/1000
S=np.random.binomial(1,r*dt,(NumTrials,len(time)))/dt
```

The resulting rate and raster plot is shown in figure 2.4c,d. Compare to the real spike trains from figure 2.2b.

Now that we understand how to model spike trains with spike timing variability, let's look at neurons driven by synapses with Poisson presynaptic spike times.

Exercise 2.2.1. Generate 10 realizations of a Poisson process with rate $r = 10$ Hz over a time interval of duration $T = 1$s and compute the sample Fano factor. Repeat this process five times with newly generated Poisson processes to see how the sample Fano factor varies over trials. Then do the same thing for 100 Poisson processes.

2.3 Modeling a Neuron with Noisy Synaptic Input

In section 1.3 in chapter 1, we modeled a postsynaptic neuron receiving input from a single excitatory and single inhibitory synapse, which was not sufficient to drive the membrane potential to threshold. We now consider a model in which a single EIF model neuron receives input from K_e excitatory and K_i inhibitory synapses (figure 2.5a). The model is defined by the following equations:

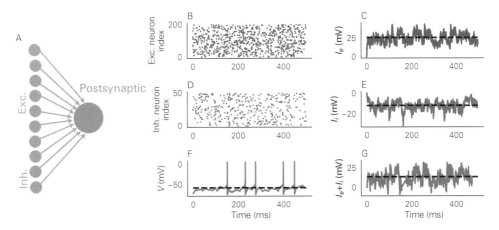

Figure 2.5
An EIF driven by synaptic input from several Poisson-spiking presynaptic neurons. (A) Schematic of the model with $K_e = 8$ excitatory and $K_i = 2$ inhibitory presynaptic neurons. (B) Raster plot of $K_e = 200$ excitatory presynaptic neurons. (C) The resulting excitatory synaptic current. Dashed line shows the stationary mean. (D,E) Same, but for $K_i = 50$ inhibitory presynaptic neurons. (F) Membrane potential of the postsynaptic neuron. Dashed line shows the mean free membrane potential. (G) Total synaptic input. See `EIFwithPoisson Synapses.ipynb` for the code to produce these plots.

EIF with input from several excitatory and inhibitory synapses

$$\tau_m \frac{dV}{dt} = -(V - E_L) + De^{(V-V_T)/D} + I_e(t) + I_i(t)$$

$$\tau_e \frac{dI_e}{dt} = -I_e + \boldsymbol{J}^e \cdot \boldsymbol{S}^e(t)$$

(2.2)

$$\tau_i \frac{dI_i}{dt} = -I_i + \boldsymbol{J}^i \cdot \boldsymbol{S}^i(t)$$

$$V(t) > V_{th} \Rightarrow \text{spike at time } t \text{ and } V(t) \leftarrow V_{re}$$

Here, $\boldsymbol{S}^e(t)$ and $\boldsymbol{S}^i(t)$ are K_e- and K_i-dimensional vectors of spike densities. We use superscripts in the place of subscripts for vectors and matrices so we can use subscripts for indices. Specifically, $\boldsymbol{S}^e_k(t)$ is the spike density of the presynaptic excitatory neuron k. Similarly, $\boldsymbol{J}^e$ and $\boldsymbol{J}^i$ are vectors of synaptic weights, so $\boldsymbol{J}^e_k$ is the synaptic weight from the presynaptic excitatory neuron k. The dot product represents the sum:

$$\boldsymbol{J}^a \cdot \boldsymbol{S}^a(t) = \sum_{k=1}^{K_a} \boldsymbol{J}^a_k \boldsymbol{S}^a_k(t), \quad a = e, i.$$

We model presynaptic spike trains as Poisson processes and generate them in the same way we generated multi-trial Poisson processes:

```
Se=np.random.binomial(1,re*dt,(Ke,len(time)))/dt
```

and similarly for `Si`. The shape of `re` determines whether all the neurons have the same or different rates, and whether the Poisson processes are homogeneous. For example, if `re` is a scalar, then all neurons will have the same rate and all spike trains will be homogeneous. To generate inhomogeneous processes, all with different rates, you'd need to use an `re` that has shape `(Ke, len(time))`. Synaptic weight vectors can also be uniform or heterogeneous. Uniform weights ($\boldsymbol{J}^e_k = j_e$) can be generated using

```
je=15.0
Je=je+np.zeros(Ke)
```

and similarly for `Ji`. Figure 2.5 shows the results with $K_e = 200$ excitatory neurons with rates $r_e = 8$ Hz and $K_i = 50$ inhibitory neurons with rates $r_i = 15$ Hz. Since a sum of Poisson processes is a Poisson process itself, only the *product* of K_a and r_a is important whenever $\boldsymbol{J}^a$ is uniform. For example, the simulation in figure 2.5 (with $K_e = 200$ and $r_e = 8$) is mathematically equivalent to one with $K_e = 100$ and $r_e = 16$ Hz, because $K_e r_e = 1600$ Hz in both cases.

The Euler step to update synaptic currents is written as

```
Ie[i+1]=Ie[i]+dt*(-Ie[i]+Je@Se[:,i])/taue
```

where @ is the NumPy symbol for a dot product or matrix multiplication. The Euler step updates to Ii and V are similar.

The membrane potential and postsynaptic spike times in figure 2.5f are noisy and qualitatively similar to the real neuron in figure 2.1a. Hence, Poisson presynaptic spike times provide a simple approach to modeling postsynaptic neural variability.

We can use the resulting mathematical model to better understand how the statistics of presynaptic spike trains map to the statistics of the postsynaptic spike train. First, consider the time-dependent expectation, $E[I_e(t)]$, which is what you would get if you simulated the model in equation (2.2) over many trials (with new random numbers on each trial), and then averaged $I_e(t)$ over the trials. To compute $E[I_e(t)]$, we first take expectations in equation (2.2) to get

$$\tau_e \frac{dE[I_e]}{dt} = -E[I_e] + E[\boldsymbol{J}^e \cdot \boldsymbol{S}^e] = -E[I_e] + \boldsymbol{J}^e \cdot \boldsymbol{r}^e$$

where $\boldsymbol{r}^e$ is the vector of firing rates and the first equality follows from equation (2.1). When $\boldsymbol{J}^e$ and $\boldsymbol{r}^e$ are uniform ($J_k^e = j_e$ and $r_k^e = r_e$, as in figure 2.5), we have

$$\boldsymbol{J}^e \cdot \boldsymbol{r}^e = K_e j_e r_e.$$

When they are not uniform the derivation can proceed without this substitution. Repeating this analysis for $I_i(t)$, we see that mean synaptic inputs obey the ordinary differential equations (ODEs):

$$\tau_e \frac{dE[I_e]}{dt} = -E[I_e] + K_e j_e r_e$$

$$\tau_i \frac{dE[I_i]}{dt} = -E[I_i] + K_i j_i r_i.$$

These equations are valid for homogeneous and inhomogeneous presynaptic spike trains ($r_e(t)$ and $r_i(t)$ time-dependent or time-constant). When r_e and r_i are time-constant, $E[I_e(t)]$ and $E[I_i(t)]$ decay exponentially to the fixed points:

$$\bar{I}_e = \lim_{t \to \infty} E[I_e(t)] = K_e j_e r_e$$

$$\bar{I}_i = \lim_{t \to \infty} E[I_i(t)] = K_i j_i r_i. \tag{2.3}$$

These are called the *stationary mean values* of $I_e(t)$ and $I_i(t)$. The expectation approaches these values after a brief transient determined by the synaptic time constants, $\tau_e, \tau_i \approx 5$ ms. In other words, if you averaged $I_e(t)$ over many trials, then the trial average at each t would converge to $\bar{I}_e$. An alternative interpretation of the stationary mean values is motivated by noticing in figure 2.5c,e that the synaptic currents (red and blue lines) tend to fluctuate around the stationary mean values (black dashed lines) after a brief transient. Along these lines, equation (2.3) can be interpreted as saying that, after a transient,

$$I_e(t) = \bar{I}_e + \text{noise}$$

$$I_i(t) = \bar{I}_i + \text{noise}$$

(2.4)

where $E[noise] = 0$. More precisely, if you take a single trial of $I_e(t)$ but average it over a long time interval, then the time-average would also converge to $\bar{I}_e$. In other words,

$$\lim_{T \to \infty} \frac{1}{T} \int_0^T I_e(t)dt = \bar{I}_e.$$

This analysis of the expectations of I_e and I_i by taking the expectations in equation (2.2) is an example of a *mean-field theory* of neural networks. The phrase "mean-field" is often used when we replace random values by their means or an approximation to their means.

The mean-field analysis given here relied on the linearity of the ODEs defining I_e and I_i in equation (2.2) (how?). The expressions that define $V(t)$ in equation (2.2) are not linear, so mean-field theory cannot be applied so easily. Indeed, there is no known closed-form expression for the postsynaptic firing rate or stationary mean membrane potential, but some approximations can be obtained. Substituting equation (2.4) into equation. (2.2) gives

$$\tau_m \frac{dV}{dt} = -(V - E_L) + De^{(V-V_T)/D} + \bar{I} + \text{noise}$$

(2.5)

where

$$\bar{I} = \bar{I}_e + \bar{I}_i$$

(2.6)

is the stationary mean input (figure 2.5g). In other words, the model in equation (2.2) behaves like an EIF driven by time-constant input with added noise.

Here's the interesting part: In the exercise at the end of section 1.2 in chapter 1, you were asked to derive the threshold input required to drive an EIF with time-constant input to spike. If you apply your result to the EIF from figure 2.5, you should get $I_{th} = 15$ mV. The neuron in figure 2.5 clearly spikes, but $\bar{I} = 12.75$ mV is subthreshold. Without the noise term in equation (2.5), the EIF wouldn't spike, but in the presence of noise, it does! This is known as *noise-driven* or *fluctuation-driven* spiking. The idea is that the stationary mean input drives the membrane potential toward threshold, but not over threshold. Then the membrane potential fluctuations (produced by noisy synaptic input) occasionally push the membrane potential over the threshold to generate spikes (Gerstein and Mandelbrot 1964; Softky and Koch 1993; Shadlen and Newsome 1998). In this way, noise can increase the firing rate of neurons.

Another way to understand noise-driven spiking is to first replace the EIF model in equation (2.2) with a leaky integrator:

$$\tau_m \frac{dV_0}{dt} = -(V_0 - E_L) + I_e(t) + I_i(t).$$

This V_0, which we get by ignoring active currents and spiking, is sometimes called the *free membrane potential*. This equation *is* linear, so we can apply the same mean-field approach that we used before to get

$$\tau_m \frac{dE[V_0]}{dt} = -(E[V_0] - E_L) + E[I_e] + E[I_i].$$

Therefore, the *stationary mean free membrane potential* is given by

$$\overline{V}_0 = \lim_{t \to \infty} E[V_0(t)] = E_L + \overline{I}$$

which is plotted as a dashed line in figure 2.5f. Hence, the dynamics of the free membrane potential look like

$$V_0(t) = \overline{V}_0 + \text{noise}.$$

In figure 2.5f, the *free* membrane potential, $V_0(t)$, would just fluctuate around $\overline{V}_0 = -59.25$ mV (the dashed line). The actual membrane potential (gray curve) also fluctuates around $\overline{V}_0$, but whenever the fluctuations drive it near $V_T = -55$ mV, they recruit the non-linear, exponential terms in the ODE for V (which are absent in the equation for V_0), and a spike can be generated. The resulting postsynaptic spike train is irregular and Poisson-like because it is driven by random fluctuations over the threshold. Hence, spike timing variability in the presynaptic spike trains drives spike timing variability in the postsynaptic spike train.

If parameters are chosen differently (*e.g.*, by increasing r_e or j_e), then we can have $\overline{I} > I_{th}$, in which case the neuron is driven to spike by the mean input; that is, the neuron would spike even if we replaced the time-varying input $I_e(t) + I_i(t)$ with its stationary mean, $\overline{I}$. However, noise still introduces some irregularity in the spike times. This is called *mean-driven* or *drift-driven* spiking. Spike timing in the drift driven regime is more regular (closer to periodic) because it is driven primarily by time-constant input, $\overline{I}$.

Figure 2.6a,b compares the membrane potentials in fluctuation- and drift-driven regimes. The black curve in figure 2.6c shows how the firing rate depends on the stationary mean synaptic input as r_e is increased. This curve, showing a neuron's firing rate as a function of its mean input, is called an *f-I curve*.

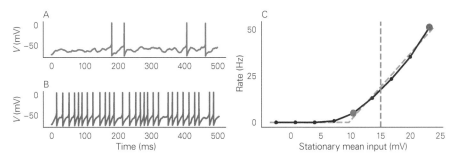

Figure 2.6
Computing an f-I curve for an EIF driven by Poisson synaptic input. (A, B) Membrane potential of an EIF in fluctuation-driven and drift-driven regimes. (C) An f-I curve for an EIF. The firing rate was plotted as a function of stationary mean synaptic input, $\overline{I}$, as r_e was varied. The blue and red dots correspond to the membrane potentials from (A) and (B). The vertical dashed line shows the cutoff between the fluctuation and drift-driven regimes (at $\overline{I} = I_{th}$). The green dashed line is a rectified-linear fit to the f-I curve. See `EIFfIcurve.ipynb` for the code to produce this figure.

The exercise at the end of section 1.2 in chapter 1 asked you to plot an f-I curve for an EIF driven by time-constant input, $I(t) = I_0$. The f-I curve in figure 2.6c shows the same thing, but for a neuron driven by Poisson synaptic input. The vertical dashed line shows the cutoff, I_{th}, between the fluctuation- and drift-driven regimes. An f-I curve for an EIF with time-constant input (such as the one from the exercise in section 1.2) would be zero to the left of the dashed line. Hence, the positivity of the f-I curve to the left of the dashed line in figure 2.6c demonstrates noise-driven spiking.

The f-I curve in figure 2.6 gives the impression that the firing rate is a function of the stationary mean input $\bar{I}$ alone, but this is not true. Two sets of parameter values that give the same value of $\bar{I}$ might produce two different firing rates. For example, if you multiply the value of j_i by 2 and multiply the value of r_i by $1/2$, then $\bar{I}$ does not change, but the postsynaptic firing rate will change. However, as exercise 2.3.2 below demonstrates, the f-I curve is *approximately* a function of $\bar{I}$ across a reasonably large range of parameters.

Motivated by these observations, we can make the approximation

$$r \approx f(\bar{I}).$$

Combining this with equations (2.3) and (2.6) gives the following:

Stationary mean-field approximation for a neuron with several synaptic inputs

$$r \approx f(w_e r_e + w_i r_i) \qquad (2.7)$$

---

where

$$w_a = K_a j_a$$

is called a *mean-field synaptic weight*, which quantifies the combined strength of all synapses from presynaptic population $a = e, i$ onto the postsynaptic neuron.

Equation (2.7) provides an *mean-field approximation* of postsynaptic firing rates. To predict the postsynaptic firing rates from a set of parameters, we only need to specify a function, f, to use as the approximate f-I curve. A simple but useful family of f-I curves are given by *rectified linear* functions (Curto, Degeratu, and Itskov 2013),

$$f(I) = (I - \theta)g H(I - \theta) = \begin{cases} (I - \theta)g & I \geq \theta \\ 0 & I < \theta \end{cases}$$

where $H(\cdot)$ is the Heaviside step function, θ is a threshold below which the firing rate is zero, and g is a *gain*, which quantifies the slope or derivative of the f-I curve when $r > 0$. In Python, we can use a curve-fitting function to fit θ and g to our simulated data:

```
from scipy.optimize import curve_fit
def f(IBar, g, theta):
    return g*(IBar-theta)*(IBar>theta)
params,_=curve_fit(f, IBars, rs)
gfit=params[0]
thetafit=params[1]
```

The dashed *green* curve in figure 2.6c shows the fit. While it's not perfect, it does provide a reasonable approximation. In chapter 3, we will use these approximations to understand the behaviour of networks of neurons. Improved approximations can be achieved by using stochastic analysis to account for the effect of presynaptic spike-timing variability on post-synaptic firing rates (Gerstein and Mandelbrot 1964; Knight 1972; Ricciardi and Sacerdote 1979; Amit and Tsodyks 1991; Lindner et al. 2004; Richardson 2007), but this approach is outside the scope of this book.

Exercise 2.3.1. Compare trial-averaged currents to time-averaged currents in simulations. Simulate a synaptic current, $I_e(t)$ (you do not need to simulate $V(t)$ or $I_i(t)$) over a fixed time interval of duration $T = 100$ ms. Repeat this in a for-loop over many trials and compute the trial-averaged value of $I_e(T)$. Compare the trial average to $\overline{I}_e = K_e J_e r_e$ as the number of trials increases. Then simulate $I_e(t)$ for one trial and compute the time average. Compare the time average to $\overline{I}_e = K_e J_e r_e$ for increasing values of T. The time average is more accurate if you throw away the transient from the first 25 ms. This is called a *burn-in period*. An estimate of an expectation obtained by averaging across trials or time is called a *Monte-Carlo estimate*. When we don't have a closed-form equation for an expectation, sometimes Monte-Carlo estimates are the best we can do.

Exercise 2.3.2. Reproduce figure 2.6 using different values of r_i, J_e, J_i or some combination thereof. However, use the fit from the original parameters in figure 2.6 to generate the dashed green line. You might need to adjust the range of r_e values to sweep out the same range of rates on the vertical axis (try for rates in $[0, 50]$ Hz). How well does the original fit capture the f-I curve under different parameters?

3

Modeling Networks of Neurons

A network is a group of neurons with some synaptic connections between them. We have already considered two networks in which one neuron receives synaptic input from all the other neurons (figures 1.5 and 2.5 in chapters 1 and 2). A network is called *recurrent* if you can follow connections to get from one neuron back to itself. Otherwise, a network is called *feedforward*. Cortical neuronal networks are recurrent, but we will begin by modeling feedforward networks because they are simpler and studying them first will help us study recurrent networks next. Network models that represent individual spikes (like those in figures 1.5 and 2.5) are called *spiking network models*. In this chapter, we begin with spiking network models and then derive *rate network models*, in which neurons' firing rates are modeled directly without representing individual spike times.

3.1 Feedforward Spiking Networks and Their Mean-Field Approximation

Feedforward networks are often arranged in *layers*, where each layer receives synaptic input from the layer before it. Let's begin by modeling a simple network with two layers (figure 3.1a). The first layer has N_e excitatory and N_i inhibitory neurons, which provide synaptic input to a second layer of N neurons.

Figure 3.1a shows two ways to sketch the network. At the top, each circle represents a neuron and arrows are individual synapses. This approach is cumbersome for large networks. The bottom illustration shows a simpler approach, in which each circle is a population and arrows indicate connected populations. An arrow between two populations does not mean that *every* pair of neurons are connected, only that *some* connections exist in the indicated direction. The network in figure 3.1a can be modeled as follows:

Feedforward spiking network model

$$\tau_m \frac{d\boldsymbol{V}}{dt} = -(\boldsymbol{V} - E_L) + De^{(\boldsymbol{V} - V_T)/D} + \boldsymbol{I}^e(t) + \boldsymbol{I}^i(t)$$

$$\tau_e \frac{d\boldsymbol{I}^e}{dt} = -\boldsymbol{I}^e + J^e \boldsymbol{S}^e$$

$$\tau_i \frac{d\boldsymbol{I}^i}{dt} = -\boldsymbol{I}^i + J^i \boldsymbol{S}^i$$

$$\boldsymbol{V}_j(t) > V_{th} \Rightarrow \text{spike at time } t \text{ and } \boldsymbol{V}_j(t) \leftarrow V_{re}$$

(3.1)

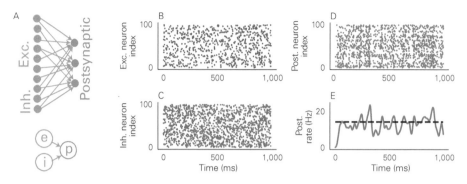

Figure 3.1
A feedforward spiking network. (A) Two ways to schematicize a feedforward network. A layer of postsynaptic neurons (gray lines) receives synaptic input from a layer of presynaptic excitatory neurons (red lines) and inhibitory neurons (blue lines). In the top schematic, each circle is a neuron. In the bottom, each circle is a population. (B, C, D) Raster plots of 100 excitatory, inhibitory, and postsynaptic neurons from a simulation with $N_e = 2,000$ excitatory, $N_i = 500$ inhibitory, and $N = 100$ postsynaptic neurons. (E) Estimated time-dependent firing rate averaged over postsynaptic neurons (gray line) and the mean-field approximation of the firing rates (dashed line). The code to reproduce this figure can be found in `FeedFwdSpikingNet.ipynb`.

Equation (3.1) is similar to equation (2.2) from chapter 2, except that the equations now have a slightly different interpretation: $V(t)$ represents an N-dimensional vector of membrane potentials, J^e is an $N \times N_e$ matrix of synaptic weights, and J^i is $N \times N_i$. These are called *connectivity matrices* or *weight matrices*. The entry J^e_{jk} represents the synaptic weight from the excitatory neuron $k = 1, \ldots N_e$ to the postsynaptic neuron $j = 1, \ldots, N$ and similarly for J^i. Note that, counter to intuition, the first index in a weight matrix refers to the postsynaptic neuron and the second index refers to the presynaptic neuron (J^e_{jk} is "from k to j," not "from j to k"). While this seems backward, it is necessary for the matrix multiplication to make sense. Specifically, the jth element of the matrix products are expanded as

$$[J^b S^b]_j = \sum_{k=1}^{N_b} J^b_{jk} S^b_k$$

for $b = e, i$ and $j = 1, \ldots, N$. In the cortex, most neurons are not synaptically connected, even if they are nearby, and connectivity appears to be partly random. A simple model of this randomness is provided by a random connection matrix defined by

$$J^b_{jk} = \begin{cases} j_b & \text{with probability } p_b \\ 0 & \text{otherwise} \end{cases}$$

where p_b is the connection probability and j_b is the synaptic weight for neurons from presynaptic population $b = e, i$. Note that $j_e > 0$ and $j_i < 0$. Connection matrices can be generated as

```
Je=je*np.random.binomial(1,pe,(N,Ne))
```

and similarly for `Ji`. As in section 2.3 of chapter 2, each presynaptic spike train, $S^e_j(t)$ and $S^i_j(t)$, can be modeled as a Poisson process with rates r_e and r_i.

The code to simulate this network model is similar to the code in `EIFwithPoisson Synapses.ipynb` from section 2.3 except for a few differences. The terms V, Ie, or Ii need to be initialized as arrays like

```
Ie=np.zeros((N,len(time)))
```

and updated like

```
Ie[:,i+1]=Ie[:,i]+dt*(-Ie[:,i]+Je@Se[:,i])/taue
```

Each individual postsynaptic neuron behaves just like the single postsynaptic neuron modeled in section 2.3, except that the number of excitatory and inhibitory synapses, K_e and K_i, received by each neuron is random instead of fixed. The *expected* numbers of excitatory and inhibitory synaptic inputs received by each postsynaptic neuron are given by $E[K_e] = p_e N_e$ and $E[K_i] = p_i N_i$, respectively. Therefore, the stationary mean-field synaptic inputs are now given by

$$\bar{I}_e = N_e p_e j_e r_e$$

$$\bar{I}_i = N_i p_i j_i r_i.$$

Mathematically,

$$\bar{I}_a = \lim_{t \to \infty} E[I_j^a(t)]$$

where the expectation is taken over randomness in $S^a(t)$ *and* randomness in J_a.

In section 2.3, we pointed out that $\bar{I}_e$ could be estimated to arbitrary precision by averaging over a sufficiently long time interval. This is not true for the randomly connected model considered here because randomness in J^e cannot be averaged out by a time average. This type of randomness is called *quenched randomness*. Instead, randomness in J^e can be averaged over postsynaptic neurons. The larger the number of postsynaptic neurons, N, the more accurate this average will become. Specifically,

$$\lim_{T,N \to \infty} \frac{1}{N} \sum_{j=1}^{N} \frac{1}{T} \int_0^T I_j^e(t)dt = \bar{I}_e.$$

As in section 2.3, we cannot derive an exact mean-field theory of postsynaptic firing rates, but we can again use an approximate f-I curve, $r \approx f(\bar{I})$, where

$$\bar{I} = \bar{I}_e + \bar{I}_i = N_e p_e j_e r_e + N_i p_i j_i r_i.$$

Putting this together gives an approximation of the stationary mean postsynaptic rates as follows:

Stationary mean-field approximation for a feedforward network

$$r = f(w_e r_e + w_i r_i) \tag{3.2}$$

---

where

$$w_a = N_a E[J^a_{jk}] = N_a p_a j_a$$

is the mean-field synaptic weight for this network model. Equation (3.2) is identical to equation (2.7) except that w_e and w_i are defined using the expected number of inputs, $E[K_a] = N_a p_a$, in place of the deterministic number of inputs, K_a, used in equation (2.7).

Figure 3.1e shows the mean-field approximation from equation (3.2) (dashed line) compared to the mean rate estimated from the full simulation (gray curve). This comparison shows that the relatively simple equation (3.2) does a decent job of describing firing rates without needing to simulate an entire network (although note that we needed to simulate the network to fit the f-I curve initially).

Equation (3.2) applies to networks with just two layers with one population in the first layer and two populations in the second, but it is easily extended to feedforward networks with arbitrary numbers of layers and populations, giving rise to equations of the following form:

Multi-layered feedforward rate network model

$$r_1 = f(W_1 r_0)$$
$$r_2 = f(W_2 r_1)$$
$$\dots$$
$$r_L = f(W_L r_{L-1})$$
$$\tag{3.3}$$

---

where r_ℓ is a vector of stationary mean firing rates in layer ℓ and W_ℓ is the mean-field connectivity matrix from layer $\ell - 1$ to layer ℓ. Equation (3.3) is often used as a model in itself (instead of an approximation to a spiking network model), in which case it is called a *feedforward rate network model*. In section 4.2 of chapter 4 and appendix B.8, we will see how equation (3.3) can be used to build artificial neural networks (ANNs) for machine learning. Multilayered networks like equation (3.3) mimic the layered architecture of the cortex. In particular, the cortex is layered in two separate senses.

First, some cortical areas are arranged in a layered, hierarchical fashion (figure 3.2a). A well-known cortical hierarchy is the ventral stream of the visual cortex. Visual stimuli from the retina pass through the thalamus to the primary visual cortex (V1), which responds to simple visual features like the orientation (angle) of edges. V1 transmits its responses to V2, and so on to reach higher visual cortical areas that respond to increasingly complex visual features like shapes and faces. Connections between cortical areas are primarily excitatory. If we visualize the cortex as a crinkled-up sheet, cortical areas are located in different regions along the surface of the sheet (figure 3.2b).

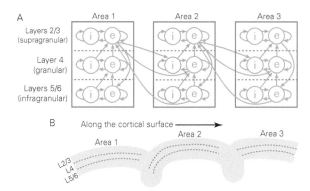

Figure 3.2
Diagram of a simplified cortical circuit model. (A) The cortex is layered in two senses. Cortical areas form layered hierarchies, and there is a stereotyped architecture of layers within each area. This diagram, adapted from the canonical architecture described in Bastos et al. (2012), shows some of the dominant connectivity pathways between cortical areas and layers. (B) Viewing the cortex as a wrinkled sheet, cortical areas are arranged along the surface of the sheet, and layers are arranged along its depth.

Second, each cortical area is composed of several *cortical layers* and these layers are connected with some stereotyped motifs (figure 3.2a). For example, cortical layer 4 typically receives input from lower cortical areas and then sends synaptic projections to layers 2/3. Cortical layers are arranged along the depth of the cortical sheet (figure 3.2b).

Despite its layered architecture, the cortex is by no means feedforward, as you can see in figure 3.2a. First, connections between cortical areas exist in both directions of the hierarchy: feedforward and feedback (*e.g.*, V1 projects to V2 and V2 projects back to V1). Second, nearby neurons within the same cortical area and cortical layer are interconnected with each other, forming local recurrent networks. We next develop recurrent spiking networks that model local recurrent connectivity within a layer.

Exercise 3.1.1. An accurate estimate of $\bar{I}_e$ can be obtained by averaging over postsynaptic neurons *and* over time. Try averaging over both in a simulation and compare the averages to $\bar{I}_e = N_a p_a j_a r_e$ for increasing values of N and/or T.

3.2   Recurrent Spiking Networks and Their Mean-Field Approximation

Despite the complexity of figure 3.2a, it is still a gross simplification of a real cortical circuit. Among other factors, there are many connectivity pathways not shown in the diagram. Moreover, within a single layer and area, there are numerous subtypes of inhibitory neurons, and connectivity depends on the neurons' subtypes and the distance between the neurons.

It would be an enormous task to account for most of these details in a single model, so we start with a much simpler model of a local patch of neurons within a single area and layer. This model is sketched out in figure 3.3a. The excitatory and inhibitory neurons connect to each other and receive synaptic input from an external population, x, representing synaptic input from different cortical layers or areas. Since connections between cortical areas are primarily excitatory, we will assume that x is an excitatory presynaptic population. The spiking model for this network is defined by equation (3.4).

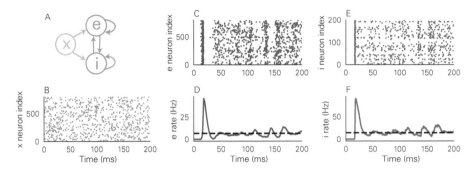

Figure 3.3
Simulation of a recurrent spiking network. (A) A schematic of the network. An external excitatory population sends synaptic input to excitatory and inhibitory populations, which are recurrently connected. (B) Raster plot of the external spike trains, which are modeled as Poisson processes. (C) Raster plot of the excitatory spike trains, which are modeled using EIF neurons. (D) Smoothed estimate of the population-averaged, time-dependent firing rate of the excitatory population. The dashed black line shows mean-field approximation to the firing rates. (E, F) Same as C and D, but for the inhibitory population. The code to reproduce this figure can be found in `RecurrentSpikingNet.ipynb`.

A recurrent spiking network model

$$\tau_m \frac{dV^e}{dt} = -(V^e - E_L) + De^{(V^e - V_T)/D} + I^{ee}(t) + I^{ei}(t) + I^{ex}(t)$$

$$\tau_m \frac{dV^i}{dt} = -(V^i - E_L) + De^{(V^i - V_T)/D} + I^{ie}(t) + I^{ii}(t) + I^{ix}(t)$$

$$\tau_b \frac{dI^{ab}}{dt} = -I^{ab} + J^{ab}S^b, \quad a = e, i, \ b = e, i, x$$
(3.4)

$$V_j^a(t) > V_{th} \Rightarrow \text{spike at time } t \text{ and } V_j^a(t) \leftarrow V_{re}, \quad a = e, i.$$

In equation (3.4), the first equation describes the N_e-dimensional vector of excitatory neurons' membrane potentials, the second equation is the same for inhibitory neurons, and the last equation defines their threshold-reset conditions. The term $I^{ab}(t)$ is the synaptic input from population b to population a. For example, $I^{ei}(t)$ is the N_e-dimensional vector of total inhibitory input to excitatory neurons. The connection matrix, J^{ab}, is an $N_a \times N_b$ matrix of synaptic weights. Spikes in the external population can be generated as Poisson processes, each with firing rate r_x. As described previously connection matrices are random:

$$J_{jk}^{ab} = \begin{cases} j_{ab} & \text{with probability } p_{ab} \\ 0 & \text{otherwise.} \end{cases}$$
(3.5)

Figure 3.3 shows spike trains from a simulation with $N_x = N_e = 800$ and $N_i = 200$. Early in the simulation, firing rates increase and then decrease quickly (figure 3.3c–f) because the initial conditions of the membrane potentials cause many neurons to cross the threshold at nearly the same time. After this transient, firing rates settle down and fluctuate around a steady state or "stationary" value.

Simulating large networks can be computationally expensive in terms of run time and memory. One source of inefficiency is the use of large arrays to represent time-varying vectors. For example, time-dependent vectors like $I^{ee}(t)$ and $V^e(t)$ can be stored as a $N_e \times N_t$ vector where $N_t = T/dt$ is the number of time bins. This type of representation appears in the `FeedFwdSpikingNet.ipynb` code used to generate figure 3.1. However, if we don't need to keep track of the history of these vectors, we can store them as N_e-dimensional vectors. In this case, an Euler step looks like

```
Iee=Iee+dt*(-Iee+Jee@Se[:,i])/taue
```

The new value of `Iee` overwrites the old value. Of course, we can do the same for all synaptic currents, $I^{ab}(t)$, and membrane potentials, $V^a(t)$. The downside to this approach is that you cannot generate plots or compute statistics of the membrane potentials or synaptic currents after the simulation, but this is not a problem if we're only interested in spike trains and firing rates, which is often the case. We generally want to keep track of the spike trains across time, so we do not use the same approach to store the spike densities $S^e(t)$, $S^i(t)$, and $S^x(t)$.

There are many more ways to increase the efficiency of large spiking network simulations. For example, since `Se`, `Si`, `Jee`, and others are sparse arrays (i.e., they contain mostly zeros), we can store and multiply them in more efficient ways. However, the number of synapses in a network of N neurons is proportional to N^2, so the time and memory required to simulate the network are also proportional to N^2. Hence, spiking network simulations are computationally expensive for large N. Since local cortical circuits contain many thousands (or even millions) of neurons, it is often useful to use mean-field equations and rate models in the place of spiking networks.

To derive a mean-field equation for stationary firing rates, we can adapt the mean-field theory that we developed for feedforward networks. The mean-field synaptic inputs to excitatory and inhibitory neurons are given by

$$\bar{I}_e = w_{ee}r_e + w_{ei}r_i + w_{ex}r_x$$
$$\bar{I}_i = w_{ie}r_e + w_{ii}r_i + w_{ix}r_x \tag{3.6}$$

where

$$w_{ab} = N_b E[J_{jk}^{ab}] = N_b p_{ab} j_{ab} \tag{3.7}$$

is a mean-field synaptic weight and r_b is the stationary firing rate of population b. We can again use the mean-field approximation, $r_a \approx f(\bar{I}_a)$, from section 3.1 to get an approximation of the form:

$$r_e = f(w_{ee}r_e + w_{ei}r_i + w_{ex}r_x)$$

$$r_i = f(w_{ie}r_e + w_{ii}r_i + w_{ix}r_x)$$

where f is an f-I curve that approximates the dependence of firing rates on inputs. It is useful to write this approximation in vector form as follows:

> ### Stationary mean-field approximation for a recurrent network
>
> $$r = f(Wr + X) \tag{3.8}$$

where

$$r = \begin{bmatrix} r_e \\ r_i \end{bmatrix}, \quad W = \begin{bmatrix} w_{ee} & w_{ei} \\ w_{ie} & w_{ii} \end{bmatrix}, \quad \text{and } X = W_x r_x = \begin{bmatrix} w_{ex} \\ w_{ix} \end{bmatrix} r_x.$$

Equation (3.8) can also be used to model networks with an arbitrary number of populations.

In contrast to equation (3.2) for feedforward networks, the unknown quantity r appears on both sides of equation (3.8). This is called an *implicit equation* because r is defined implicitly as a solution. Recurrent networks produce *implicit* mean-field equations, and feedforward networks produce *explicit* mean-field equations. In general, equation (3.8) can have one solution, many solutions, or no solutions for r.

For some choices of f, we can find explicit solutions to equation (3.8). If we use the threshold-linear f-I curve, $f(I) = (I - \theta)gH(I - \theta)$ from section 2.3 of chapter 2, we can look for solutions in which $r_e, r_i > 0$. Any such solution satisfies the linear equation

$$r = (Wr + X - \theta)g$$

which has the explicit solution

$$r = [I - gW]^{-1}(X - \theta)g \tag{3.9}$$

where I is the identity matrix. This solution is plotted as dashed lines in figure 3.3d,f. Despite its simplicity, equation (3.9) provides a reasonable approximation to firing rates in the spiking network.

Equation (3.8) describes stationary firing rates, but it does not describe the stability of these rates or the intrinsic dynamics of the network. To describe stability and intrinsic dynamics, we need to derive a *dynamical mean-field equation*, which is a system of ordinary differential equations (ODEs) that approximates the dynamics of mean firing rates. Different derivations lead to different mean-field formulations. We relegate those details to appendix B.5 and focus here on a particular formulation, given by equation (3.10).

> ### Dynamical mean-field approximation for a recurrent network
>
> $$\tau_e \frac{dr_e}{dt} = -r_e + f(w_{ee}r_e + w_{ei}r_i + X_e)$$
>
> $$\tau_i \frac{dr_i}{dt} = -r_i + f(w_{ie}r_e + w_{ii}r_i + X_i). \tag{3.10}$$

In equation (3.10), τ_e and τ_i no longer represent synaptic time constants; instead, they represent the combined timescales of synaptic dynamics and membrane potential dynamics (see appendix B.5 for more details). Generally, they should be in the range of 10–50 ms and should be larger for populations with slower synaptic dynamics *or* membrane dynamics.

Equation (3.10) is easily extended to networks with several populations, and it can be interpreted as a model in itself, which is the topic of the next section. Equation (3.10) is very similar to a mean-field model called the "Wilson-Cowan equations," which were first proposed by Hugh Wilson and Jack Cowan for modeling interacting excitatory and inhibitory populations (Wilson and Cowan 1972, 1973).

> See appendix A.8 for a review of fixed points and stability in systems of ODEs.

Fixed points of equation (3.10) are given by equation (3.8), so the two equations are consistent. The stability of the fixed points is determined by the Jacobian matrix (see appendix A.8):

$$J = \begin{bmatrix} (g_e w_{ee} - 1)/\tau_e & g_e w_{ei}/\tau_e \\ g_i w_{ie}/\tau_i & (g_i w_{ii} - 1)/\tau_i \end{bmatrix}$$

where $g_a = f'(w_{ae}r_e + w_{ai}r_i + w_{ax}r_x)$ is called the *gain* of population a, which quantifies the sensitivity of the rate to input perturbations. If all the eigenvalues of J have negative real parts at a particular fixed point, then that fixed point is stable. If any eigenvalues have positive real parts, it is unstable.

Figure 3.4a,b shows a simulation of equation (3.10) using parameters from the spiking network simulation in figure 3.3. Checking the eigenvalues shows that the fixed point is stable, which is confirmed numerically by the convergence of the firing rates toward the fixed points (dashed lines). The steady state rates in figure 3.4a,b are close to those in figure 3.3, but the dynamics during the relaxation to steady state are very different. This is because the dynamical mean-field equations do not capture the effect where many neurons reach the threshold at the same time in the spiking model. Regardless, equation (3.10) can help us understand instabilities in the spiking model, as we describe next.

Oscillations in systems of ODEs can emerge through a Hopf bifurcation where a pair of complex eigenvalues changes from having a negative to a positive real part (see

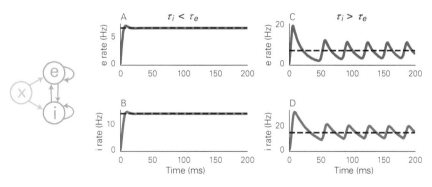

Figure 3.4
Simulation of dynamical mean-field equations for a recurrent network. (A, B) Excitatory (red) and inhibitory (blue) firing rates from a simulation of equation (3.28) with $\tau_e = 30$ ms and $\tau_i = 15$ ms. Rates quickly approach their fixed point (dashed lines), indicating stability. (C, D) Same simulation with $\tau_e = 15$ ms and $\tau_i = 30$ ms. Rates oscillate around their fixed points, indicating a Hopf bifurcation. Parameters were chosen to match the mean-field theory from figure 3.3, so the dashed lines are in the same place. The code to reproduce this figure can be found in `DynamicalMeanField.ipynb`.

appendix A.8). In two-dimensional systems like equation (3.10), this happens when the trace of the Jacobian matrix changes from negative to positive while the determinant remains positive. For equation (3.10), the trace is given by

$$Tr(J) = \frac{g_e w_{ee} - 1}{\tau_e} + \frac{g_i w_{ii} - 1}{\tau_i} \tag{3.11}$$

and recall that stability requires $Tr(J) < 0$. Note that $w_{ii} < 0$, so the only positive contribution to the trace is the $g_e w_{ee}$ term. When $g_e w_{ee} < 1$, the trace is always negative. Excitatory-inhibitory networks with stable fixed points satisfying

$$g_e w_{ee} > 1$$

are called *inhibitory-stabilized networks (ISNs)* because the network would be unstable without an inhibitory population, so inhibition stabilizes the network (Tsodyks et al. 1997; Ozeki et al. 2009). You can check that the network would be unstable without inhibition by drawing a phase line for the dynamics of the excitatory population by itself. Let's assume that the network satisfies $g_e w_{ee} > 1$ (which is the case for the network in figure 3.4a,b). Then stability requires that the inhibitory term in equation (3.11) is large enough in magnitude to cancel the excitatory term:

$$\left| \frac{g_i w_{ii} - 1}{\tau_i} \right| > \left| \frac{g_e w_{ee} - 1}{\tau_e} \right|.$$

Let's focus on the impact of the time constants here. Stability is encouraged by fast inhibition and comparatively slower excitation (τ_i smaller or τ_e larger, or both). If inhibition becomes too slow compared to excitation (τ_i too large compared to τ_e), the trace will become positive, producing oscillations through a Hopf bifurcation. In summary, if the system in equation (3.10) is in the inhibitory-stabilized regime, sufficiently slow inhibition or fast excitation will give rise to oscillations. Indeed, figure 3.4c,d demonstrates oscillations that arise when we swap the values of τ_e and τ_i.

The oscillations can be understood as follows: Recurrent excitation is strong enough to produce a runaway positive feedback loop, but inhibition shuts down this loop before it can blow up. This is the defining property of an ISN. If inhibition is fast enough, it shuts down the runaway excitation before it can grow at all. This is the stable condition. If inhibition is too slow, the runaway excitation starts taking off before inhibition can shut it down, and then the process starts over again, producing an oscillation. This is called a PING oscillation since it arises from the interaction between Pyramidal (excitatory) and INhibitory neurons (or INterneurons). PING oscillation are believed to be a source of fast (Gamma) oscillations in the brain (Kopell et al. 2000; Whittington et al. 2000; Brunel and Wang 2003; Cardin et al. 2009).

ISNs predict a surprising phenomenon, sometimes called the *paradoxical effect*. This paradoxical effect occurs when a small *increase* in external input to inhibitory neurons causes inhibitory firing rates to *decrease*. In other words, increasing X_i by a small amount leads to a decrease in the fixed point of r_i. Paradoxical effects have been observed in cortical recordings, and it can be shown that they arise in the model from equation (3.10) only when the network is in an inhibitory stabilized state (Tsodyks et al. 1997; Ozeki et al. 2009; Sanzeni et al. 2020). In the next section, we use this fact to understand phenomena observed in real cortical recordings.

Exercise 3.2.1. Let's test whether slow inhibition or fast excitation produces oscillations in a spiking network. Repeat the simulation from figure 3.4, but slow the inhibitory synapses by taking $\tau_i = 8$ ms. You should see strong oscillations emerge.

Computational models in the literature (including many of my own papers) often take inhibitory synapses to be faster than excitation to avoid strong oscillations. However, in real cortical circuits, inhibition is a little bit slower than the fastest form of excitation (AMPA-ergic synapses). This raises the question of how the cortex avoids strong oscillations. Some hypotheses have been proposed (Huang et al. 2019; Ahmadian and Miller 2021), but the full answer to this question is not known.

Exercise 3.2.2. Starting from the model in figure 3.4a,b, increase X_i by a small amount and test whether a paradoxical effect occurs. Now decrease w_{ee} until the network is no longer an ISN and repeat the test. You can also try this in the recurrent spiking network model from figure 3.3.

3.3 Modeling Surround Suppression with Rate Network Models

The mean-field equations described in the previous section did a good job of approximating steady state mean firing rates, but they did not capture the dynamics of the rates while they approach their steady states (compare figure 3.3 to figure 3.4a). Similar, they captured the general phenomenon in which slower excitation causes oscillations (compare figure 3.4b to exercise 3.2.1), but they cannot accurately predict the values of τ_e and τ_i at which oscillations occur or the frequency and amplitude of oscillations in spiking networks.

This mixed success might seem damning for rate models, but recall that spiking models are already a major simplification of real neural circuits. When using simplified models of complex systems, it is not always practical to seek *quantitatively* accurate predictions. In other words, we shouldn't expect our model to tell us whether firing rates in a real biological neural circuit are closer to 10 Hz or 15 Hz. Instead, simplified models should provide insight into general principles and qualitative phenomena, like the presence of oscillations when inhibition is too slow compared to excitation.

To this end, if we care only about understanding phenomena related to firing rates, then detailed spiking network models are unnecessary. We can skip the details here and just model firing rates directly. The dynamical mean-field in equation (3.10) is a good place to start. Generalizing equation (3.10) to an arbitrary number of populations gives equation (3.12).

Dynamical rate network model

$$\boldsymbol{\tau} \circ \frac{d\boldsymbol{r}}{dt} = -\boldsymbol{r} + f(W\boldsymbol{r} + \boldsymbol{X}(t)). \tag{3.12}$$

Equation (3.12) models the dynamics of M firing rates stored in the M-dimensional vector $\boldsymbol{r}$. The term $\boldsymbol{\tau}$ is an M-dimensional vector of time constants. The $\circ$ represents an element wise product or "Hadamard product" between two vectors. If $\boldsymbol{z} = \boldsymbol{x} \circ \boldsymbol{y}$, then $z_k = x_k y_k$. In other words, $\circ$ performs the same operation as $\star$ in NumPy. In equation (3.12), this means that each term in the vector can have its own time constant. If a single time constant is chosen, then we can use a scalar τ and omit the $\circ$.

Equation (3.10) is a special case of equation (3.12) and fixed points are given by equation (3.8). The stability of a fixed point is determined by the eigenvalues of the Jacobian matrix:

$$J = \frac{1}{\tau} \circ [-I + GW]$$

where I is the identity matrix and G is a diagonal matrix with entries given by the gains, $G_{kk} = g_k = f'(I_k)$, where $I = Wr + X$ is the input at the fixed point. Equation (3.12) can be used to model a feedforward network by setting $W = 0$, but the dynamics of feedforward rate networks are uninteresting (they just decay exponentially to their fixed points), so static rate models (like equations (3.2) and (3.3)) are more commonly used for feedforward networks.

Equation (3.12) can have two different interpretations. Each element of r can be interpreted as the mean firing rate of *a population of neurons* as in equation (3.10), or each element can be interpreted as *an individual neuron*. In other words, equation (3.12) can model a network with M populations or a network with M neurons.

An alternative formulation of rate networks uses a system of ODEs for the synaptic inputs instead of the rates:

$$\tau \frac{dI}{dt} = -I + Wf(I + X) \tag{3.13}$$

where firing rates are given by $r = f(I + X)$. Equation (3.13) is discussed more in appendix B.5, but we will focus on the formulation in equation (3.12) here.

Dynamical rate network models (as discussed in this section) are mathematically equivalent to dynamical mean-field equations (as discussed in the previous section). The difference is that, in this section, we do not attempt to relate the rate equations to spiking network models, but instead interpret them as models in themselves. This leads to a much simpler modeling approach since equation (3.12) has far fewer parameters than a spiking network model. We can further simplify the choice of parameters by nondimensionalizing some of them. We still want to interpret τ to have the unit in *ms* and r to have the unit in kHz = (1/ms), but we can interpret W to be dimensionless (which implies that X and $I = Wr + X$ have the units in kHz). We can also simplify the definition of the f-I curve to

$$f(I) = [I]^+ = IH(I) = \begin{cases} I & I > 0 \\ 0 & I < 0 \end{cases}$$

which is a dimensionless version of the rectified linear f-I curve from section 2.3 in chapter 2. With this convention, every term in equation (3.12) is in ms or 1/ ms = kHz, or is dimensionless. Hence, rate networks allow us to model firing rates without explicitly quantifying any biophysical units like mV for membrane potentials.

To demonstrate how rate network models make it easy to model cortical phenomena, we next consider the phenomenon of *surround suppression*. As mentioned in section 2.1, individual neurons in the primary visual cortex (V1) increase their firing rates in response to visual stimuli inside a small region of the visual field, called the neuron's receptive field. A drifting grating stimulus at the neuron's preferred orientation that lies entirely inside the neuron's receptive field is called a *center stimulus*, and it will typically increase the neuron's firing rate (figure 3.5a, first half of the red curve). The region of the visual field just outside a neuron's receptive field is called its *surround*. Interestingly, a larger stimulus that lies partly outside the neuron's visual field (a "center+surround" stimulus) will often evoke a

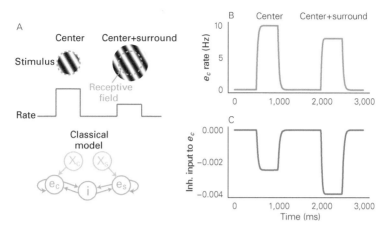

Figure 3.5
A simple model of surround suppression. (A) Illustration of surround suppression. The center stimulus is presented inside a neuron's visual field, which causes an increase in the neuron's firing rate. When a surround stimulus is added outside the neuron's visual field, the neuron's firing rate is suppressed. A simple network model of surround suppression has separate center and surround excitation populations coupled together through a shared inhibitory population. (B) The model captures the suppression of the center neuron's firing rates. (C) The model predicts an increase in inhibition to the center neurons when they are suppressed. The code to reproduce this figure can be found in the first code cell of `SurroundSuppression.ipynb`.

lower firing rate than a center stimulus (figure 3.5a, second half of the red curve). In other words, neurons' firing rates are suppressed by stimuli in their surrounds (Allman, Miezin, and McGuinness 1985). These results show that the neurons' firing rate is affected by stimuli outside its receptive fields, so the term "classical receptive field" is sometimes used in place of "receptive field" to explicitly exclude the surround.

What causes surround suppression? Computational models of surround suppression are abundant, but they almost all come down to one explanation: *One neuron's surround is another neuron's center.* When the surround stimulus is turned on, it elevates the firing rates of neurons whose classical receptive fields overlap with the surround. These elevated rates can increase the inhibition and/or decrease the excitation to the neurons whose classical receptive fields contain the center stimulus.

A simple model of surround suppression has two excitatory populations, e_c and e_s, which do not connect directly to each other but form reciprocal connections to the same inhibitory population (figure 3.5a, bottom). Population e_c models neurons with receptive fields in the, center and population e_s models neurons with receptive fields in the surround. Connectivity is given by

$$r = \begin{bmatrix} r_{e_c} \\ r_{e_s} \\ r_i \end{bmatrix} \text{ and } W = \begin{bmatrix} w_{ee} & 0 & w_{ei} \\ 0 & w_{ee} & w_{ei} \\ w_{ie} & w_{ie} & 0 \end{bmatrix}.$$

The external input is modeled by

$$X = \begin{bmatrix} X_{e_c} \\ X_{e_s} \\ 0 \end{bmatrix} = \begin{bmatrix} CX_e \\ SX_e \\ 0 \end{bmatrix}$$

where $C = 1$ and $S = 0$ corresponds to the center-only stimulus and $C = S = 1$ corresponds to the center+surround stimulus. Excluding the time constants (which are not important for capturing the fixed point effects), this model has only four free parameters. A spiking network model would have many more parameters to choose.

Figure 3.5b shows firing rates from a simulation of the model. When the surround stimulus is turned on, firing rates in the center population decrease. Examining the network diagram in figure 3.5a, it is easy to see why the suppression occurs: Increased excitation to e_s (by turning on X_{e_s}) causes r_{e_s} to increase, which increases r_i, thereby increasing the inhibition to e_c and suppressing its firing rates. This intuition is supported by observing that the surround stimulus increases inhibitory input to population e_c (figure 3.5c). This general phenomenon in which two neural populations mutually inhibit each other is sometimes referred to as *competition* between the populations. The network in figure 3.5a is the simplest model of competition between two excitatory populations.

In 2009, this simple model of surround suppression was challenged by recordings in cat V1 collected by the neuroscientists Hirofumi Ozeki and Ian Finn, working in the laboratory of David Ferster. Their recordings showed that the surround stimulus causes a *decrease* in inhibition to neurons tuned to the center stimulus. The decreased inhibition is paired with a decrease in excitation, which together leads to a net decrease in firing rates. The Ferster lab teamed up with the computational neuroscientists Evan Schaffer and Kenneth Miller to model and understand these observations (Ozeki et al. 2009).

The simplest model that they developed included just one excitatory and one inhibitory population, both representing neurons tuned to the center stimulus (figure 3.6a). The surround stimulus was modeled by increased input to the inhibitory population. This surround stimulus represents extra input to inhibitory neurons in the presence of a surround stimulus, and it can model synaptic input from the neurons in the same layer, a different layer, or a different cortical area. Putting this together gives equation (3.12) with

$$W = \begin{bmatrix} w_{ee} & w_{ei} \\ w_{ie} & w_{ii} \end{bmatrix} \text{ and } X = \begin{bmatrix} CX_e \\ CX_i + SX_i \end{bmatrix}.$$

Ozeki et al. (2009) proved that this model exhibits surround suppression with an increase in inhibition to the excitatory neurons exactly when the network is in an inhibitory stabilized state. This conclusion is related to the paradoxical effect described at the end of section 3.2 since the extra input to the inhibitory population *decreases* inhibitory rates, leading to decreased inhibition to the excitatory population. Therefore, not only does the model capture the decrease in inhibition during surround suppression, but it also supports the conclusion that cat V1 is in an inhibitory stabilized state. Figure 3.5b,c shows the results from a simulation of the model in an inhibitory stabilized state. Note that the surround stimulus suppresses r_e while decreasing inhibition to population e.

Both of the models discussed here are major simplifications of the actual circuit dynamics underlying surround suppression. The first model inexplicably omits inhibitory-to-inhibitory connections (w_{ii}) and connections between e_c and e_s. The second does not model a surround population at all, and it assumes no surround-evoked change in external input to e neurons. Many of these details can be added without changing the basic conclusions reached with the models (Ozeki et al. 2009).

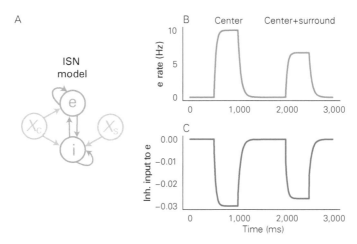

Figure 3.6
An alternative model of surround suppression. (A) Illustration of the model of surround suppression from Ozeki et al. (2009). (B) The model captures the suppression of the center neuron's firing rates. (C) When the model is in an inhibitory stabilized state, it also captures the decrease in inhibition to excitatory neurons during suppression. The code to reproduce this figure can be found in the second code cell of `SurroundSuppression.ipynb`.

A more detailed model might have four populations: e_c, i_c, e_s, and i_s, with connections between all pairs of populations (so sixteen weights in W). Moreover, there are at least three subtypes of inhibitory neurons that differ in their connectivity properties (Pfeffer et al. 2013) and their roles in surround suppression (Adesnik et al. 2012), so a more complete model might have as many as eight populations with sixty-four weights in W.

Ultimately, the model chosen to describe a phenomenon like surround suppression should be as simple as possible, but detailed enough to capture the questions being asked and to make testable predictions. The two-population model considered here is very simple, but it is sufficient to capture the suppression of inhibitory currents and make predictions about the transient dynamics of inhibition. Overall, rate network models are much simpler than spiking models, and they are often sufficient for modeling most rate-based phenomenon like surround suppression.

Exercise 3.3.1.   In the inhibitory stabilized state, the model from Ozeki et al. (2009) predicts that the surround stimulus suppresses the excitatory firing rates while increasing the inhibition that they receive. What does the model predict when it is not in the inhibitory stabilized state? Rerun the simulation from figure 3.6, but replace W with cW, where $c < 1$ is small enough that the network is no longer in the inhibitory stabilized state.

Exercise 3.3.2.   We have focused on steady state firing rates, but the model in Ozeki et al. (2009) also makes a prediction about transient rate dynamics. In particular, if the surround stimulus is turned on while the center stimulus is already on (*i.e.*, a direct transition from center to "center+surround), then the inhibitory input to excitatory neurons should transiently increase before it decreases. This effect was observed in recordings from cat V1 (Ozeki et al. 2009). Modify the simulation from figure 3.5 to verify that this occurs when the network is in an inhibitory stabilized state. Then check whether it occurs when the network is not inhibitory stabilized.

4

Modeling Plasticity and Learning

All of the models considered so far in this book don't really *do* anything, in the sense that they don't learn or solve any problems. The primary purpose of the brain is presumably to *do* things, or at least to tell the body to do things. Hence, all the models considered so far arguably ignore the central purpose of the brain. This does not necessarily mean that the models are useless. They can be useful for understanding and interpreting recorded neural data, and they can be viewed as building blocks from which we can build models that actually "do something."

That said, there is an argument to be made that we should focus, at least partly, on models that *can* do things. If we were modeling an electric motor, we would likely want a model that rotates a rotor. Likewise, if we are modeling the brain, perhaps we should use models that actually learn some task, even a simple one. In this chapter, we start building models that can learn and perform simple tasks. To begin with, we need to understand how to model synaptic plasticity, which is a primary mechanism of learning in the brain.

4.1 Synaptic Plasticity

In neural circuits, the strength of synaptic connections changes over time, an effect known as *synaptic plasticity*. An increase in synaptic strength is called *facilitation*, and a decrease is called *depression*. Synapses can change strength transiently for a duration of milliseconds or seconds, which is called *short-term plasticity*. We will not discuss short-term plasticity in this book, but instead focus on *long-term plasticity*, in which changes to synaptic strengths are static. Long-term plasticity is widely believed to be the primary mechanism behind learning and memory in the brain, but the precise mechanics of how learning and memory emerge from plasticity are not fully understood.

We will not go into the biophysical details of how and why synapses change strength, but many instances of plasticity are at least partially caused by an influx of calcium into a neuron after an action potential. As such, changes to the strength of a synapse can depend on the spike times and firing rates of the presynaptic and postsynaptic neurons.

The most well known type of plasticity is *Hebbian plasticity* (named after the Canadian psychologist Donald Hebb), in which the increase in synaptic strength is proportional to the firing rates of presynaptic and postsynaptic neurons (Hebb 1949). Hebbian plasticity is often described using the idiom "Neurons that fire together wire together." In other words, if a pair of presynaptic and postsynaptic neurons tend to spike nearby in time, or increase their firing rates at the same time, then the synapse between them tends to get stronger. In appendix B.6, we describe how a form of Hebbian plasticity can give rise to assembly formation and associative memory in *Hopfield network models*.

For a the dynamical rate network model from equation (3.12), pure Hebbian plasticity on the weight W_{jk} can be defined by

$$\frac{dW_{jk}}{dt} = c \boldsymbol{r}_j \boldsymbol{r}_k.$$

If we impose plasticity on all weights, this can be written as

$$\frac{dW}{dt} = c \boldsymbol{r} \boldsymbol{r}^T$$

where c is a constant. The problem with this form of pure Hebbian plasticity is that it can be unstable. Since W_{jk} can only increase, it has a tendency to grow without bound. This effect is amplified by a positive feedback loop: If W_{jk} increases, it causes r_j to increase, which causes W_{jk} to increase, and so on. There are many biologically motivated approaches for resolving this instability. We will discuss a form of Hebbian-like plasticity that is self-stabilizing.

In particular, we will build and analyze a model of *homeostatic inhibitory synaptic plasticity* in which inhibitory synapses onto excitatory neurons are modulated in a way that pushes the excitatory firing rates toward a stable target rate (Castillo, Chiu, and Carroll 2011; Vogels et al. 2011, 2013; Luz and Shamir 2012; Hennequin, Agnes, and Vogels 2017; Schulz et al. 2021; Capogna, Castillo, and Maffei 2021). First, consider a single excitatory neuron receiving synaptic input from a single inhibitory neuron with synaptic strength $J_{ei} < 0$ (figure 4.1a). The rule is defined by

$$\frac{dJ_{ei}}{dt} = -\epsilon \left[(y_e - 2r_0) S_i - S_e y_i \right]$$

$$\tau_y \frac{dy_e}{dt} = -y_e + S_e \qquad\qquad (4.1)$$

$$\tau_y \frac{dy_i}{dt} = -y_i + S_i.$$

Here, $S_e(t)$ and $S_i(t)$ are spike densities for the excitatory and inhibitory spike trains, $J_{ei} < 0$ is the synaptic weight from the i neuron to the e neuron, $r_0 > 0$ is a parameter called the *target rate* (for reasons we'll see soon), and $\epsilon > 0$ is a *learning rate* that controls how quickly the synaptic weight changes. The terms $y_e(t)$ and $y_i(t)$ are called *eligibility traces*, and they serve as continuous time estimates of the neurons' recent firing rates since $y_a(t) = k * S_a(t)$ with $k(t)$ an exponential kernel; see chapters 1 and 2. The timescale, τ_y, of the eligibility traces is related to the timescale of intracellular calcium and should be taken to be around $\tau_y \approx 100 - 1{,}000$ ms.

The first line in equation (4.1) in chapter 4 is the plasticity rule itself. Let's interpret what it is saying. Since $S_e(t)$ and $S_i(t)$ are sums of Dirac delta functions, J_{ei} is updated only after a spike in one of the neurons. After each inhibitory spike, J_{ei} is decremented by $y_e(t) - 2r_0$.

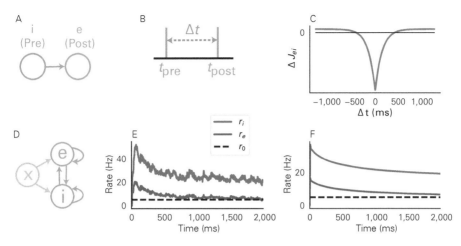

Figure 4.1
Inhibitory synaptic plasticity in a pair of neurons and a network. (A) Diagram of a single inhibitory synapse onto an excitatory neuron. (B) Each neuron spikes once, and $\Delta t = t_{post} - t_{pre}$ is the delay between spikes. (C) The change in synaptic weight, ΔJ_{ei}, as a function of Δt. (D,E) A recurrent spiking network like the one in figure 3.3 except J^{ei} evolves through synaptic plasticity, which pushes excitatory firing rates (red line) toward their target (black dashed line). (F) A mean-field rate network model approximated the firing rate dynamics from the spiking network. The code to reproduce this figure can be found in `SynapticPlasticity.ipynb`.

After each excitatory spike, it is incremented by $y_i(t)$. Note that $J_{ei} < 0$ since it is inhibitory, so decrementing it makes the synapse stronger (i.e., more inhibitory).

Plasticity rules that depend on the timing of presynaptic and/or postsynaptic spikes are called *spike-timing dependent plasticity (STDP)* rules. They are widely observed in cortical networks. To understand the dependence of the weight change on spike timing, let's consider how the synaptic weight changes in response to a single spike in each neuron (figure 4.1b). Suppose that the inhibitory neuron (which is presynaptic) spikes at time $t_{pre} > 0$, and the excitatory neuron (which is postsynaptic) spikes at time $t_{post} > 0$. Further, assume that those are the only two spikes in recent history, so we can set initial conditions $y_e(0) = y_i(0) = 0$.

Let's first consider what happens when the inhibitory (presynaptic) neuron spikes first ($t_{pre} < t_{post}$). All terms are zero before the presynaptic spike time. In particular, $y_e(t_{pre}) = 0$, so at the time of the presynaptic spike, the synaptic weight will change by

$$\Delta J_{ei} = 2\epsilon r_0.$$

Next, the excitatory (postsynaptic) neuron spikes and causes an additional change, given by

$$\Delta J_{ei} = -\epsilon y_i(t_{post}).$$

Now, note that $y_i(t)$ is defined by an exponential decay after the inhibitory (presynaptic) spike time, so

$$y_i(t_{post}) = \frac{1}{\tau_y} e^{-(t_{post} - t_{pre})/\tau_y}.$$

Taken together, the total change caused by the two spikes is given by

$$\Delta J_{ei} = -\frac{\epsilon}{\tau_y} \left(e^{-(t_{post} - t_{pre})/\tau_y} - 2\tau_y r_0 \right).$$

Now, let's compute the weight change when the excitatory (postsynaptic) neuron spikes first. By similar reasoning, we have

$$\Delta J_{ei} = -\frac{\epsilon}{\tau_y} \left(e^{-(t_{pre}-t_{post})/\tau_y} - 2\tau_y r_0 \right).$$

Putting this all together gives an unconditional weight change of

$$\Delta J_{ei} = -\frac{\epsilon}{\tau_y} \left(e^{-|\Delta t|/\tau_y} - 2\tau_y r_0 \right)$$

where

$$\Delta t = t_{post} - t_{pre}$$

is the time elapsed between the two spikes. This curve is plotted in figure 4.1c. Synaptic plasticity is often measured in experiments using the *paired pulse protocol*, in which a pair of synaptically connected neurons is driven to spike consecutively. The change in synaptic strength is computed (averaged across many trials), and plots like figure 4.1c are created for real neurons.

This paired pulse approach for quantifying the synaptic plasticity is not ideal. Under natural settings, pairs of spikes in presynaptic and postsynaptic neurons do not occur in isolation, but the neurons are spiking continuously in response to inputs from other neurons. To capture this more realistic scenario, we next consider a recurrent spiking network (figure 4.1d) with inhibitory plasticity, defined by

$$\tau_m \frac{dV^e}{dt} = -(V^e - E_L) + De^{(V^e - V_T)/D} + I^{ee}(t) + I^{ei}(t) + I^{ex}(t)$$

$$\tau_m \frac{dV^i}{dt} = -(V^i - E_L) + De^{(V^i - V_T)/D} + I^{ie}(t) + I^{ii}(t) + I^{ix}(t)$$

$$\tau_b \frac{dI^{ab}}{dt} = -I^{ab} + J^{ab}S^b, \quad a=e,i, \quad b=e,i,x \tag{4.2}$$

$$V_j^a(t) > V_{th} \Rightarrow \text{spike at time } t \text{ and } V_j^a(t) \leftarrow V_{re}, \quad a=e,i$$

$$\tau_y \frac{dy^a}{dt} = -y^a + S^a \quad a=e,i$$

$$\frac{dJ^{ei}}{dt} = -\epsilon \left[(y^e - 2r_0)[S^i]^T - S^e[y^i]^T \right] \circ \Omega^{ei}.$$

Equation (4.2) is the most complicated model that we will consider in this book. The first four equations are the same as in section 3.2 in chapter 3, and the last two implement the inhibitory plasticity rule at the network level. The notation $[S^i]^T$ and $[y^i]^T$ denoted the transpose of each vector. The matrix Ω_{ei} is just a binarized version of J^{ei},

$$\Omega_{jk}^{ei} = \begin{cases} 1 & J_{jk}^{ei}(0) \neq 0 \\ 0 & J_{jk}^{ei}(0) = 0 \end{cases}$$

where $J^{ei}(0)$ is the initial value of the matrix J^{ei}. Recall that $\circ$ denotes elementwise multiplication. Hence, the inclusion of Ω^{ei} in equation (4.2) makes sure that the plasticity rule is only applied to connected neurons and the strength of connection between unconnected neurons remains zero.

Note that elements in J^{ei} can become positive from this plasticity rule, which is not biologically realistic. If we want to prevent this, we can add the following line to our code:

```
Jei=np.minimum(Jei,0)
```

after updating `Jei`. See `SynapticPlasticity.ipynb` for a full implementation of the model in equation (4.2). Figure 4.1e shows firing rates from a simulation. Note that the excitatory firing rates (red) seem to approach the target rate, r_0 (dashed black). We next use a mean-field analysis to explain this result.

To derive a dynamical mean-field model of the spiking network simulation in equation (4.2), we can first use the rate network model from equation (3.10) to model firing rate dynamics. The entire matrix, J^{ei}, is reduced to a single mean-field weight, w_{ei}, in this approximation. Now we need to use the last equation in equation (4.2) to derive a mean-field approximation to the dynamics of w_{ei}. Recall from section 3.1 of chapter 3 that the relationship between w_{ei} and J^{ei} should be

$$w_{ei} = E[J^{ei}]N_i.$$

As in our previous mean-field derivations, we replace the spike densities, S^a, in equation (4.2) with rates, r_a. Since eligibility traces, y_a, are just running estimates of the rates, we also replace y_a with r_a. Putting all this together with the last equation in equation (4.2) gives

$$\frac{dw_{ei}}{dt} = -\epsilon \left[(r_e - 2r_0)r_i + r_e r_i \right] p_{ei} N_i.$$

Simplifying this expression and putting it into a mean-field rate model gives equation (4.3).

A rate network model with homeostatic inhibitory plasticity

$$\tau_e \frac{dr_e}{dt} = -r_e + f_e(w_{ee}r_e + w_{ei}r_i + X_e)$$

$$\tau_i \frac{dr_i}{dt} = -r_i + f_e(w_{ie}r_e + w_{ii}r_i + X_i) \qquad (4.3)$$

$$\frac{dw_{ei}}{dt} = -\epsilon_r(r_e - r_0)r_i$$

---

where

$$\epsilon_r = 2\epsilon p_{ei} N_i$$

is a rescaled learning rate, $X_e = w_{ex}r_x$, and $X_i = w_{ix}r_x$. The last equation in equation (4.3) shows that the inhibitory plasticity rule is almost like a pure Hebbian rule except for the subtraction of r_0 from r_e.

Figure 4.1f shows simulations of this rate model, which capture the overall trends seen in the spiking network. More important, equation (4.3) helps us understand why excitatory rates approach r_0: Any fixed point of the system in equation (4.3) must satisfy $r_e = r_0$ (unless $r_i = 0$, which is not the case here). Hence, the inhibitory synaptic plasticity rule pushes excitatory firing rates toward the target rate, r_0.

Exercise 4.1.1. The fixed-point analysis given here applies only for time-constant input $X_e(t) = X_e$ and $X_i(t) = X_i$. When external input changes in time, the network cannot generally maintain a fixed firing rate $r_e = r_0$. Try running a rate network simulation in which external input changes in time. For example, try a simulation where $X_e(t)$ and/or $X_i(t)$ change their values halfway through the simulation or change at periodic intervals.

4.2 Feedforward Artificial Neural Networks

We began this chapter by saying that we will build models that learn to "do something," but the model in the previous section hardly fulfills that promise because it only learns to produce a constant target rate, which is not a very useful task. In this section, we build a model that can learn to solve real tasks. As our models become more capable of solving real tasks, they will also become more abstract and removed from biology.

You may have heard of *deep neural networks (DNNs)*, which are a leading tool for machine learning and artificial intelligence. Even if you haven't heard of them, you've very likely used them. If you've ever translated a webpage online, used facial recognition or voice recognition software, or employed many of the other seemingly magical modern applications of artificial intelligence, then you have very likely benefited from the power of DNNs. What is the word "neural" doing in the term "deep neural networks"? The development of DNNS and their predecessors were inspired by biological neural networks. Indeed, DNNs are types of artificial neural networks (ANNs), which are closely related to rate network models as we explain next.

Let us start by considering a *feedforward, single-layer, fully connected ANN*, which is equivalent to a feedforward mean-field equation with one presynaptic and one postsynaptic layer. In particular, the ANN can be defined by removing recurrent connections (setting $W = 0$) in equation (3.8) to get $\boldsymbol{r} = f(W_x r_x)$. However, we will switch notational conventions to write equation (4.4):

> **Single layer ANN**
>
> $$\boldsymbol{v} = f(W\boldsymbol{x}) \tag{4.4}$$

which can also be written as $\boldsymbol{v} = f(\boldsymbol{z})$, where

$$\boldsymbol{z} = W\boldsymbol{x}.$$

This gives us a mapping from a vector of inputs, $\boldsymbol{x} \in \mathbb{R}^{N_0}$ to outputs, $\boldsymbol{v} \in \mathbb{R}^{N_1}$. The outputs are also sometimes called *activations*, $W \in \mathbb{R}^{N_1 \times N_0}$ is the *feedforward weight matrix*, and $f : \mathbb{R} \to \mathbb{R}$ is called the *activation function*, which is applied pointwise—that is, $\boldsymbol{v}_j = f(\boldsymbol{z}_j)$. The term "fully connected" refers to the fact that W connects every element of $\boldsymbol{x}$ to every element of $\boldsymbol{z}$ or $\boldsymbol{v}$.

In *supervised learning*, we begin with a data set of m inputs and labels:

$$\{x^i, y^i\}_{i=1}^m$$

The goal in supervised learning is to find a weight matrix, W, such that the relationship between x and v in equation (4.4) approximates the relationship between x^i and y^i from the data.

To measure how well the current value of W performs, we use a *loss function*, $L(v, y)$, which measures some notion of distance or error between v and y. We can then quantify the performance of our network by a *cost function*, which averages the loss over the entire data set:

$$J(W) = \frac{1}{m} \sum_{i=1}^m L(v^i, y^i)$$

where $v^i = f(Wx^i)$ is the output of the network on input x^i. Sometimes the word "loss" is used to refer to the cost, but it's usually easy to tell which is meant from the context.

For a specific example of supervised learning, we consider the MNIST data set (LeCun et al. 1998), which consists of 28×28 grayscale images of handwritten digits, which are the inputs. The full MNIST data set contains 70,000 images, but we will restrict ourselves to a subset of $m = 1{,}000$ images. We represent the inputs as vectors in $N_0 = 28 * 28 = 784$ dimensions; that is, each input is a list of pixel values. The labels are 10-dimensional binary vectors with a 1 in the entry associated with the digit. This is called *one-hot encoding*. For example, the one hot-encoded label for a handwritten 2 is

$$y = [0\ 0\ 1\ 0\ 0\ 0\ 0\ 0\ 0\ 0]^T$$

where the first entry corresponds to digit 0, so 2 is the third digit. Outputs of the network, therefore, should have dimension $N_1 = 10$ and W should be 10×784. Figure 4.2a shows a diagram of the network. Since the entries of the labels are in the interval $[0, 1]$, we should use an activation function that returns outputs in $[0, 1]$. We will use the logistic sigmoid function (figure 4.2b):

$$f(z) = \sigma(z) = \frac{e^z}{e^z + 1}.$$

This is an S-shaped function, which is strictly increasing and has horizontal asymptotes at 0 and 1. S-shaped functions like $\sigma(z)$ are called *sigmoidal* functions. One nice property of the logistic sigmoid is that its derivative can be written in terms of the function itself:

$$\sigma'(z) = \sigma(z)(1 - \sigma(z)).$$

We will use a *mean squared error (MSE)* loss function:

$$L(v, y) = \frac{1}{2} \|v - y\|^2 = \frac{1}{2} \sum_{j=1}^{10} (v_j - y_j)^2 \tag{4.5}$$

which is proportional to the squared Euclidean distance between v and y. Technically, we should divide by 10 instead of 2 to get a "mean" squared error, but the difference is not important, and the coefficient of $1/2$ makes the math work out nicely later.

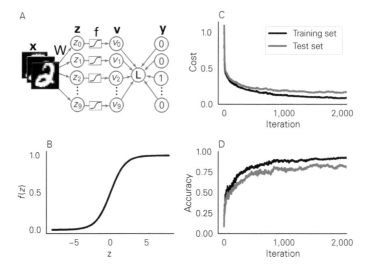

Figure 4.2
A single-layer ANN trained on the MNIST classification. (A) Diagram of an ANN with images as input and one hot-encoded labels. (B) The logistic sigmoid activation function. (C,D) The cost and accuracy of the training and test data as the model learns, expressed as a function of the number of gradient descent iterations. Accuracy is defined as the proportion of inputs classified correctly. The code to reproduce this figure can be found in `SingleLayerANN.ipynb`.

Our goal is to find a matrix, W, that achieves a small cost. The idea behind *learning* or *training* W is to start with some initial value of W, compute the loss or cost using that value of W, and then update W in such a way that the loss or cost tends to decrease. We then repeat this procedure iteratively. To learn effectively, we need to estimate how changing each term, W_{jk}, affects the cost. This is called *credit assignment* because we need to assign "credit" to each weight for its effect on the loss.

For small changes to the weights, we can use calculus to perform credit assignment. Suppose that we compute an output, $v = f(Wx)$, and a loss, $L(v, y)$, for one data point using some initial matrix, W. Then consider a small update:

$$W_{new} = W_{old} + \Delta W \text{ where } \Delta W = \epsilon U.$$

Here, U is a matrix and $\epsilon > 0$ is a learning rate. A Taylor expansion of $L(v, y)$ to first order in ϵ gives

$$L_{new} = L_{old} + \epsilon \sum_{j,k} U_{jk} \frac{\partial L}{\partial W_{jk}} + \mathcal{O}(\epsilon^2)$$

where $\mathcal{O}(\epsilon^2)$ denotes terms that go to zero quadratically as $\epsilon \to 0$. From this equation, we can see that choosing

$$U_{jk} = -\frac{\partial L}{\partial W_{jk}}.$$

will cause all the terms in the sum here to be negative (except where they are zero). Therefore, when $\epsilon > 0$ is sufficiently small, if we set $\Delta W_{jk} = -\epsilon \frac{\partial L}{\partial W_{jk}}$, then the loss will decrease (unless all the partial derivatives are zero). This choice of update is called *gradient descent*, and it can be written more concisely as

$$\Delta W = -\epsilon \nabla_W L \qquad (4.6)$$

where $\nabla_W L$ is the *gradient* of $L(\boldsymbol{v}, \boldsymbol{y})$ with respect to W, which is a matrix with entries given by $[\nabla_W L]_{jk} = \partial L / \partial W_{jk}$. Gradient descent follows the direction of steepest descent in the sense that, for small ϵ, no other update with the same magnitude can decrease the loss by more.

For our single-layer ANN, the gradient descent update is given by

$$\Delta W = -\epsilon [\nabla_{\boldsymbol{v}} L(\boldsymbol{v}, \boldsymbol{y})] \circ f'(\boldsymbol{z}) \boldsymbol{x}^T$$

where $\circ$ is elementwise multiplication, $\boldsymbol{x}^T$ is the transpose of $\boldsymbol{x}$, $\nabla_{\boldsymbol{v}} L(\boldsymbol{v}, \boldsymbol{y})$ is the gradient of the loss with respect to $\boldsymbol{v}$, and $f'(\boldsymbol{z})$ is the derivative of f applied pointwise to $\boldsymbol{z}$. For the MSE loss from equation (4.5), $\nabla_{\boldsymbol{v}} L(\boldsymbol{v}, \boldsymbol{y}) = \boldsymbol{v} - \boldsymbol{y}$ and the resulting update rule is known as *the delta rule* (Widrow and Hoff 1960):

The delta rule

$$\Delta W = -\epsilon(\boldsymbol{v} - \boldsymbol{y}) \circ f'(\boldsymbol{z}) \boldsymbol{x}^T. \qquad (4.7)$$

In *online gradient descent*, we iteratively apply the update in equation (4.6) for each of the m data points, $\boldsymbol{x}^i$ and $\boldsymbol{y}^i$. Each update to W is called one *iteration*, and each loop through the whole data set (*i.e.*, m iterations) is called one *epoch*. In NumPy, online gradient descent with MSE loss (*i.e.*, using the delta rule) can be implemented as follows:

```
for k in range(NumEpochs):
    for i in range(m):
        z=W@X[:,i]   # compute inputs
        v=f(z)     # compute activations
        DeltaW=-epsilon*np.outer((v-Y[:,i])*fprime(Z),X[:,i])
        W=W+DeltaW
```

where the ith input and label are stored as `X[:,i]` and `Y[:,i]`, respectively. The function, `outer`, takes the outer product, $\boldsymbol{x}\boldsymbol{y}^T$, between two vectors.

We could instead update W based on the gradient of the cost function by choosing

$$\Delta W = -\epsilon \nabla_W J(W) = -\frac{\epsilon}{m} \sum_{i=1}^{m} \nabla_W L(\boldsymbol{v}^i, \boldsymbol{y}^i).$$

This procedure, which is sometimes called *full-batch gradient descent*, is different from online gradient descent because it updates W using the average gradient over the entire data set instead of updating it once for each data point. Alternatively, we could average the gradient over a random subset of the data points for each update, which is called *stochastic gradient descent (SGD)*. SGD is the most common learning algorithm used to train neural networks for machine learning applications. Online gradient descent is arguably more similar to biological learning and synaptic plasticity because updates to W are computed based on the current activity of the network, without needing to store a history of gradients as we iterate through the data set.

Figure 4.2c shows the loss of our single-layer ANN on MNIST trained with online gradient descent. It's difficult to interpret performance looking at the loss alone. Instead, we can define the network's best "guess" on an input by the value of j at which v_j is the largest. We can then ask whether the network was correct—that is, whether the guess matches the true label. The *accuracy* of the network is then defined as the proportion of inputs on which the network's guess is correct. Figure 4.2d shows the accuracy during training.

So far, we checked the network's accuracy only on the data set on which it was trained. This is cheating in a sense because the network has already seen the training data, and it could just memorize those inputs. Typically, the true goal of learning is to perform well on unseen inputs. In `SingleLayerANN.ipynb`, we also checked the performance on a separate set of data that was hidden from the network during training, which is called the *test data* or *validation data*. In this context, the data used to train the model is called the *training data*. The ability to perform well on unseen data is called *generalization*. The model performed similarly on the test and training data (figure 4.2c,d), but slightly worse on the test data, as expected. Animals are excellent at generalizing, and ANNs strive to approximate this ability.

Exercise 4.2.1.   *This is my favorite exercise in the book. It ties everything together!*
Let's create a more biologically relevant task than MNIST. In section 2.1 in chapter 2, we looked at how the orientation of drifting grating stimulus is encoded in the spike count of a real neuron in a monkey's brain. Appendix B.4 extends this analysis to populations of neurons using statistical approaches. Let's now train a single-layer ANN (which is a biologically inspired model) that takes spike trains recorded from a monkey's brain as input and learns to classify the orientation of the drifting grating that the monkey is watching. This can be viewed as a model of how the spiking activity of recorded neurons might be decoded by a downstream population of neurons (modeled by the ANN) that computes what the monkey is seeing. The code in `DecodeSpikeCountsWithSingleLayerANN.ipynb` loads a matrix, X, of spike counts from $N_0 = 112$ neurons in a monkey's visual cortex recorded over $m = 1,000$ trials. On each trial, the monkey viewed a drifting grating stimulus with one of $N_1 = 12$ orientations. The matrix Y contains the one-hot encoded orientations for each trial.

Train a single-layer ANN to predict the orientation from the spike counts. Test your trained model on the test data in `XTest` and `YTest`.

After you complete this exercise, you will have built an algorithm that can *read a monkey's brain!* This approach can be extended to build brain-machine interfaces for robotic arms, artificial eyes, and other functions.

Extensions of single-layer ANNs and their relationship to biological neural networks.   In this book, we drew a continuous thread of models from the membrane dynamics of single neurons to the single-layer ANN developed in this section. While this continuum suggests a relationship between biological neural networks and ANNs, there are many caveats and missing pieces that make it difficult to take this relationship literally.

For example, the ANN from equation (4.4) is equivalent to the feedforward mean-field rate models discussed in section 3.1 in chapter 3. However, as discussed at the end of that section, cortical circuits are not feedforward. Specifically, neurons in the same layer are recurrently interconnected. Hence, a more realistic ANN would be given by a *recurrent* rate model like the one in equation (3.12). Recurrent ANNs *are* used in machine learning, but the learning rules used to update recurrent weights are more complicated and it is not clear how they could be implemented by biologically realistic synaptic plasticity rules (Lillicrap and Santoro 2019). An alternative approach called *reservoir computing* leaves recurrent weights fixed and trains a readout matrix from the recurrent network (Jaeger 2001; Maass, Natschläger, and Markram 2002; Jaeger 2004; Lukoševičius and Jaeger 2009; Sussillo and Abbott 2009) using a learning rule that is very similar to the Delta rule from equation (4.7). This approach, which is more widely used in computational neuroscience than machine learning, is described in appendix B.7.

Even if we accept the use of a feedforward architecture, the ANN model in equation (4.4) has only one layer, whereas cortical circuits are hierarchical (see the end of section 3.1), so it would be more biologically realistic to use a multilayer feedforward network like the one from equation (3.3). Multilayer feedforward networks, called *deep neural networks (DNNs)*, are powerful machine learning models. However, credit assignment in DNNs is more difficult than in single-layer ANNs. It is typically achieved using an algorithm called *backpropagation*, which can be interpreted as propagating gradients or "errors" backward in the network. It is still unknown whether or how the brain might implement or approximate backpropagation, and this is currently a highly active area of research (Whittington and Bogacz 2019; Lillicrap et al. 2020). Appendix B.8 discusses backpropagation and the issues with its potential approximation in biological neural circuits.

Aside from the architectural differences mentioned in this chapter ANNs are very different from biological neural circuits at a finer-grained level too. For example, ANNs don't have action potentials, don't have separate excitatory and inhibitory neurons (i.e., they don't obey Dale's law), and the learning rules used to train them are very different from spike timing-dependent plasticity rules.

Notably, the fact that ANNs are equivalent to rate network models, not spiking network models, shows that they ignore the details of spike timing. If spike timing beyond firing rates is important for any aspects of neural computation in the brain, then ANNs do not capture these aspects. Many neuroscientists question whether the brain would throw away all the potential information and computational power of spike times by using only firing rates for coding and computation.

Nevertheless, ANNs are becoming among the most popular and successful class of models for studying neural circuits. There are two perspectives for using ANNs as models of biological neural circuits.

One perspective is that the finer details shouldn't matter. Representations of stimuli learned by ANNs should approximate those learned by real brains even if the details of their implementation are very different. This perspective implicitly relies on some form of universality; that is, that different learning algorithms and models trained on similar tasks will have similar representations. This perspective has some support: Networks designed for machine learning produce activations that are correlated with neural activity recorded from animals viewing the same inputs as stimuli (Yamins et al. 2014; Khaligh-Razavi and Kriegeskorte 2014; Kell et al. 2018; Schrimpf et al. 2020).

Another perspective is that ANNs *do* approximate the details of biological networks to some extent (via their relationship to rate models, for example), and we should be able to obtain more accurate models by making ANNs operate more like biological neural networks (Richards et al. 2019). For example, we could impose Dale's law on ANNs

(Cornford et al. 2020), train the networks with biologically plausible plasticity rules, or train spiking networks instead of rate networks (Guerguiev, Lillicrap, and Richards 2017; Huh and Sejnowski 2018; Neftci, Mostafa, and Zenke 2019; Bellec et al. 2020; Li et al. 2021; Payeur et al. 2021).

Each perspective likely has some merit. There are probably some biological details that are important when modeling learning in neural circuits and some biological details that do not matter much. Which details are important likely depends on which data are being modeled, which questions are being asked, or both.

If we care about modeling neural activity or behaviour only in a *trained* animal, then using biologically realistic learning rules might not be important. If we care about neural activity or behaviour *during learning*, then it might be more important to use biologically realistic learning rules. If we are interested in the role of different ion channels or neurotransmitter molecules in learning, then we should use models that let us account for those details.

As in many applications of computational and mathematical modeling, a primary challenge is the choice and design of the model, which is more of an art than a science. The model needs to be chosen in such a way that it can be expected to capture the phenomena being modeled, answer the questions being asked, and make experimentally testable predictions. But the model should be as simple as possible under those constraints. The following relevant quotation is often attributed to Einstein. It is probably apocryphal, but it paraphrases some of his actual writing (Robinson 2018): "Everything should be made as simple as possible, but no simpler."

This advice should serve as a guide for designing models of neural circuits and for designing mathematical models of natural phenomena more generally.

Appendix A: Mathematical Background

A.1   Introduction to ODEs

A *differential equation* is an equation that relates an unknown function to its derivatives. For example, consider the equation

$$\frac{dy}{dt} = f(y, t) \tag{A.1}$$

where f is a known function, $y : \mathbb{R} \to \mathbb{R}$ is an unknown function, and $dy/dt = y'(t)$ is the derivative of y. The notation $y : \mathbb{R} \to \mathbb{R}$ means that y is a function that takes a real number as input and returns a real number. *Ordinary differential equations (ODEs)* like equation (A.1) are sometimes also written using the notation

$$y' = f(y, t) \quad \text{or} \quad \dot{y} = f(y, t).$$

All these equations mean the same thing: We are looking for a function $y(t)$ satisfying

$$y'(t) = f(y(t), t)$$

for all values of t in the domain being considered. Of course, y, t, and f can have different names too, like $u'(x) = g(u, x)$, $V'(t) = f(V, t)$, or $x'(t) = h(x, t)$, and the meaning is the same.

Equation (A.1) is an ODE because $y(t)$ depends on only one variable, t, so the equation contains only "ordinary" derivatives, not partial derivatives like $\partial y(x, t)/\partial x$ or $\partial y(x, t)/\partial t$, which give rise to partial differential equations (PDEs). All differential equations considered in this book are ODEs. Equation (A.1) is a *first-order ODE* because it contains only first-order derivatives, not higher derivatives like $y''(t)$. We will only consider first-order equations in this book.

Equation (A.1) is a *one-variable* ODE because it contains only one unknown function, $y(t)$, which is a scalar, not a vector. First-order ODEs in one-variable are sometimes called one-dimensional ODEs or ODEs in one dimension.

We will also consider *ODEs in higher dimensions*, which are sometimes called *systems of ODEs*. They can be written in vector form, like

$$\frac{d\boldsymbol{u}}{dt} = \boldsymbol{F}(\boldsymbol{u}, t)$$

where $\boldsymbol{u} : \mathbb{R} \to \mathbb{R}^n$, meaning that $\boldsymbol{u}$ is a function that takes a real number as input and returns an n-dimensional vector. Similarly, $\boldsymbol{F} : \mathbb{R}^n \to \mathbb{R}^n$, meaning that $\boldsymbol{F}$ takes and returns a vector, which are denoted by boldface. Systems of ODEs can also be written as a list of n one-dimensional equations like

$$\frac{dx}{dt} = f(x, y, t)$$

$$\frac{dy}{dt} = g(x, y, t)$$

where $x : \mathbb{R} \to \mathbb{R}$ and $y : \mathbb{R} \to \mathbb{R}$. This is a system of two ODEs, or a two-dimensional system. We can write this in vector form by defining the vector

$$\boldsymbol{u}(t) = \left[\begin{array}{c} x(t) \\ y(t) \end{array} \right] \in \mathbb{R}^2.$$

We will initially focus on ODEs in one dimension, but will consider ODEs in higher dimensions in section A.8.

ODEs like the one in equation (A.1) are typically paired with *initial conditions* of the form

$$y(t_0) = y_0$$

which specifies the value at some time t_0. An ODE paired with an initial condition is called an *initial value problem (IVP)*:

$$\frac{dy}{dt} = f(y, t)$$

$$y(t_0) = y_0. \tag{A.2}$$

Equation (A.2) should be interpreted to mean that we are looking for a function, $y(t)$, satisfying

$$y'(t) = f(y(t), t) \text{ and } y(t_0) = y_0.$$

In many cases, we will take $t_0 = 0$, so the IVP will be written as

$$\frac{dy}{dt} = f(y, t)$$

$$y(0) = y_0.$$

If $f(y, t)$ is continuous in t and $\partial f / \partial y$ is a continuous and bounded function of y and t, then there is a unique solution, $y(t)$, to the IVP in equation (A.2), at least over some interval containing t. "Uniqueness" means that there is only one function satisfying the IVP. The assumptions on $f(y)$ can be relaxed to some extent, and, in many cases, the IVP will have a unique solution for all $t \in (-\infty, \infty)$. See Hirsch, Smale, and Devaney (2012) or another textbook on ODEs for a more in-depth discussion of the existence and uniqueness of solutions to ODEs.

In this book, we will forgo the theory of existence and uniqueness and just always assume that f is sufficiently good that there is a unique solution to the IVP under consideration over the time interval under consideration.

For some ODEs, we can write down *closed-form* solutions. A closed-form solution is a solution for $y(t)$ that can be written in terms of known functions. This approach of finding a closed-form solution can work well for some simple ODEs. For example, this is the approach that we take for linear ODEs in sections A.2 and A.4. However, for many ODEs, we cannot write a closed-form solution. For example, try to find a closed-form solution to

$$\frac{dy}{dt} = e^{\cos(y)} + t$$

$$y(0) = 0.$$

You will not succeed. If we cannot find a closed-form solution, there are two alternative approaches that we can use instead:

1. Find a numerical approximation to the solution. This approach (which is used in section A.5) requires plugging in specific numbers for any parameters that define the ODE and initial value.

2. Analyze properties of solutions without actually solving the equation. This approach (which is used in section A.6 later in this appendix) can sometimes be more informative than a numerical solution because it can help you understand what happens for a variety of initial conditions, parameters, or both.

A.2 Exponential Decay as a Linear, Autonomous ODE

We begin by considering an IVP of the form

$$\tau \frac{dx}{dt} = -x$$

$$x(0) = x_0$$

(A.3)

where $x : \mathbb{R} \rightarrow \mathbb{R}$ is a scalar function and $\tau > 0$ is a scalar parameter. This ODE is *autonomous* because the right side, $f(x, t) = -x$, does not depend on t; and it is *linear* because the right side is a linear function of x. This is one of the simplest differential equations, but understanding this equation can help to understand more complicated equations that come up in a lot of applications. You can check that the solution is given by

$$x(t) = x_0 e^{-t/\tau}.$$

(A.4)

Equation (A.4) is the quintessential example of *exponential decay*: As time progresses, $x(t)$ decays exponentially toward zero. See the blue curve in figure A.1 for a plot of a solution and ExponentialDecay.ipynb for code to produce this plot.

Note that equation (A.4) is a solution to equation (A.3) for all $t \in (-\infty, \infty)$, not just $t > 0$. Therefore, equation (A.5) tells us about the future *and* the past values of $x(t)$.

The parameter, τ, is called the *time constant* of the decay, and it determines how quickly the decay occurs. Larger τ means slower decay and vice versa. This can be seen by computing the proportion by which x changes over a time window of duration τ, starting at some time $t = t_1$:

$$\frac{x(t_1 + \tau)}{x(t_1)} = \frac{x_0 e^{-(t_1 + \tau)/\tau}}{x_0 e^{-t_1/\tau}} = e^{-1} \approx 0.37.$$

In other words, every τ units of time, x is multiplied by a factor of about 0.37. If you make the ballpark approximation $0.37 \approx 0.5$, then τ is a rough estimate of the half-life of

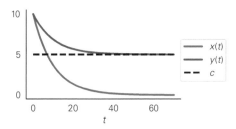

Figure A.1
Exponential decay. The exponentially decaying functions, $x(t)$ and $y(t)$, as defined by equations (A.4) and (A.6). Parameters are $\tau = 10$ and $c = 5$. See `ExponentialDecay.ipynb` for the code to produce this plot.

x, defined as the time it takes for x to be reduced by half. The true half-life is actually closer to

$$t_{half} \approx 0.7\tau.$$

So the true half-life is a little shorter than τ, but τ gives a ballpark approximation that's easier to remember and compute in your head. To see an illustration of this conclusion, note that the blue curve in figure A.1 reaches the halfway point, $x(t) = 5$, just before time $t = \tau = 10$.

Equation (A.3) can be used to model processes that decay exponentially toward zero, but we often want to model processes that decay exponentially to other values. This is easily achieved by setting

$$y(t) = x(t) + c$$

which decays exponentially to c since $x(t)$ decays to zero. What differential equation does $y(t)$ obey? This can be derived by computing

$$\tau\frac{dy}{dt} = \frac{dx}{dt} = -x = -(y-c) = -y+c$$

where we used the assumption that $y = x + c$. This tells us that the IVP

$$\tau\frac{dy}{dt} = -y+c$$

$$y(0) = y_0$$

(A.5)

should give solutions that decay exponentially to c. Indeed, you can check that the solution to equation (A.5) is given by

$$y(t) = c + (y_0 - c)e^{-t/\tau}.$$

(A.6)

While equation (A.6) looks more complicated than equation (A.4), it is easy to verify that it just represents exponential decay toward c starting at $y(0) = y_0$. The half-life now represents the time to get halfway to c instead of halfway to 0. See the red curve in figure A.1 for a plot of a solution.

Sometimes initial conditions are given at some nonzero time; that is,

$$\tau\frac{dy}{dt} = -y+c$$

$$y(t_0) = y_0.$$

(A.7)

It is easy to check that the solution to equation (A.7) is just given by sliding (i.e., translating) equation (A.6) over by t_0 to get

$$y(t) = c + (y_0 - c)e^{-(t-t_0)/\tau} \tag{A.8}$$

which also represents exponential decay toward c. Note that equation (A.8) is equivalent to equation (A.6) if we take $t_0 = 0$, and equivalent to equation (A.4) if we also take $c = 0$, so equation (A.8) is the most general form of exponential decay.

The next step will be to replace the c in equation (A.7) with a term that depends on time. First, though, we need to take a detour to introduce convolutions.

Exercise A.2.1. Derive an exact equation for the true half-life from equation (A.3) and verify that it is approximately equal to 0.7τ.

A.3 Convolutions

Given two scalar functions, $f, g : \mathbb{R} \to \mathbb{R}$, their *convolution* is another scalar function denoted $(f * g)(t)$ defined by

$$(f * g)(t) = \int_{-\infty}^{\infty} f(s)g(t-s)ds.$$

If f or g has a bounded domain, we compute the integral assuming that they are zero outside their domain. Convolution is commutative:

$$(f * g)(t) = (g * f)(t) = \int_{-\infty}^{\infty} g(s)f(t-s)ds$$

so you can use either of the two integrals given here to represent the convolution. Convolution is also linear in each argument in the sense that

$$((f + g) * h)(t) = (f * h)(t) + (g * h)(t)$$

and

$$((cf) * g)(t) = c(f * g)(t)$$

for scalars, $c \in \mathbb{R}$.

Convolutions arise in many contexts. For example, if X and Y are independent continuous random variables with probability density functions f_X and f_Y, then the probability density function of $Z = X + Y$ is $f_Z = f_X * f_Y$. The solutions to many ODEs and PDEs can be written as convolutions. Convolutions are also used in signal processing, for example, to smooth data. A two-dimensional version of convolutions, defined for functions $f, g : \mathbb{R}^2 \to \mathbb{R}$, is widely used for image processing and machine learning.

In this textbook, we focus on the convolution of a *signal*, $x(t)$, with a *kernel*, $k(t)$, which is sometimes called a *filter*. The absolute value of the kernel should have a finite integral:

$$\int_{-\infty}^{\infty} |k(s)|ds < \infty \tag{A.9}$$

and should satisfy

$$\lim_{s \to \pm\infty} k(s) = 0.$$

Therefore, we normally think of a kernel as some kind of "bump," often centered at $s = 0$, such as $k(s) = e^{-s^2}$. The signal, $x(t)$, does not need to have a finite integral or converge to zero as $t \to \pm\infty$, but it must be bounded above and below:

$$\max_t |x(t)| < \infty. \tag{A.10}$$

More precisely, there needs to be a finite number M for which $|x(t)| < M$ for all t. Unlike the kernel, the signal can be a time series that fluctuates indefinitely, such as $x(t) = \sin(t)$. The result of the convolution is given by

$$y(t) = (k * x)(t) = \int_{-\infty}^{\infty} x(s)k(t - s)ds. \tag{A.11}$$

When equations (A.9) and (A.10) are satisfied, a theorem known as *Young's inequality* tells us that $\max_t |y(t)| < \infty$. In other words, the convolution of a kernel with a signal produces another signal. What does this new signal, $y(t)$, represent in terms of the kernel and the original signal, $x(t)$?

To get an intuition, let's think about the value of the convolution for some particular time, $t = t_0$. We have

$$y(t_0) = \int_{-\infty}^{\infty} x(s)k(t_0 - s)ds.$$

Now, consider define the flipped version of the kernel: $k^-(s) = k(-s)$. In other words, $k^-(s)$ is just $k(s)$ with the time axis flipped (see figure A.2a). The convolution can be written in terms of the flipped kernel as

$$y(t_0) = \int_{-\infty}^{\infty} x(s)k^-(s - t_0)ds.$$

Treated as a function of s, note that $k^-(s - t_0)$ is just the function $k^-(s)$, shifted to the right by an amount t_0 (see figure A.2a). To get $y(t_0)$, we take $k^-(s - t_0)$, multiply it by an unshifted $x(s)$, and then integrate the product. This operation is illustrated in figure A.2b.

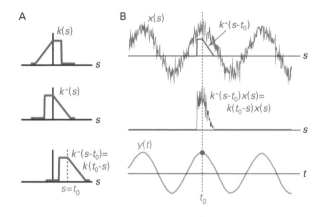

Figure A.2
Illustration of a convolution. (A) A kernel, $k(s)$, is flipped to get $k^-(s) = k(-s)$, and then shifted by t_0 to get $k^-(s - t_0) = k(t_0 - s)$. (B) A signal, $x(s)$ (blue), is multiplied by the flipped and shifted kernel, $k^-(s - t_0)$ (red). The product (purple curve) is integrated to get the value of $y(t_0)$ (purple dot). Repeating this process for all values of $t = t_0$ gives the full curve, $y(t)$ (gray).

In summary, when performing a convolution at $t = t_0$, we are taking the integral of $x(s)$ weighted by the flipped and shifted kernel, $k^-(s - t_0)$. If we think of $k(s)$ as a bump centered at $s = 0$, then the convolution is given by flipping the bump, shifting it sideways, and taking the integral of $x(s)$ weighted by the flipped and shifted kernel.

If we also assume that $k(s) \geq 0$ and

$$\int_{-\infty}^{\infty} k(s)ds = 1$$

then the integral that defines $y(t_0)$ represents a *sliding, weighted average* of $x(s)$ near $s = t_0$. The shape of the bump represented by $k(s)$ determines how we weight the values of $x(s)$ near $s = t_0$ when computing the weighted average. If the bump is wide and decays slowly, then we include values of $x(s)$ that are far from $s = t_0$ in our weighted average. If the bump is very narrow, then we only include values very close to $s = t_0$.

If $k(s) = 0$ for $s < 0$–that is, the bump represented by $k(s)$ lies solely on the positive time axis–then $k(t_0 - s) = 0$ whenever $s > t_0$, so we can write $y(t_0)$ as

$$y(t_0) = \int_{-\infty}^{\infty} x(s)k(t_0 - s)ds = \int_{-\infty}^{t_0} x(s)k(t_0 - s)ds.$$

This implies that the value of $y(t_0)$ only depends on values of s with $s < t_0$. In other words, $y(t)$ only depends on the past values of $x(t)$. For this reason, kernels for which $k(s) = 0$ for $s < 0$ are called *causal kernels* or, more commonly, *causal filters*.

In applications of convolutions, it is rare that the integral defining the convolution can be computed by hand. Instead, we typically approximate it numerically. If we wanted to use standard numerical integration, this could be a computationally expensive task because we need to approximate a new integral for each value of t. If we discretized time into 1,000 bins, we would need to perform numerical integration 1,000 times.

Fortunately, numerical approximations to convolutions can be computed very efficiently using an algorithm called the Fast Fourier Transform (FFT). More precisely, the FFT can be used to very efficiently perform a discrete-time convolution, which is a convolution defined on discrete series in which the continuous-time functions, $x(t)$ and $k(t)$, in equation (A.11) are replaced by discrete-time sequences, x_n and k_n, and the integrals are replaced by sums. A discrete-time convolution can be used to represent a Riemann approximation to the integrals in equation (A.11). Therefore, a Riemann approximation to continuous-time convolutions can be computed efficiently using the FFT.

In Python, discrete time convolutions are implemented using the built-in NumPy function, `convolve`. The code snippet that follows convolves a noisy signal with a Gaussian kernel to smooth the noise. The results are shown in figure A.3.

```python
# Discretized time
T=50; dt=.1; time=np.arange(0,T,dt)
# Define a noisy signal
x=np.sin(2*np.pi*time/20)+.25*np.random.randn(len(time))
# Define a Gaussian kernel and normalize it
sigma=2
s=np.arange(-3*sigma,3*sigma,dt)
```

```
k=np.exp(-(s**2)/sigma**2)
k=k/(np.sum(k)*dt)
# Perform convolution
y=np.convolve(x,k,'same')*dt
```

Let's go through this block of code step by step. There are a few caveats for defining a kernel:

• Recall that convolution involves sliding the flipped kernel along the signal. For this to work well, the kernel must be much shorter than the signal; otherwise, there's no room for sliding. To see this better, visualize sliding the red kernel along the blue signal in figure A.2b. If the kernel were as wide as the signal, there would be no room to slide it. This is why we define the kernel over a much shorter discretized time vector than `time`.

• Due to the way that `np.convolve` is defined, the numerical convolution approximates the integrals defining the convolution only if the kernel is centered at $t = 0$. This is why the discretized time vector s must be centered at 0.

• The time vector, s, should be wide enough to capture all of the values of k that are not close to zero. Since k is a Gaussian with width parameter `sigma`, using an interval of radius `3*sigma` is sufficient.

• If we want the convolution to represent an average, then the integral of the kernel should be 1. This is why we divided k by the Riemann approximation to its integral in the line `k=k/(np.sum(k)*dt)`. Of course, there might also be instances where we don't want the convolution to represent an average, so we would not need this line.

The last line of code performs the convolution to get the resulting signal, y. There are a few things to note about this line too:

• The option `'same'` tells `convolve` that the output, y, should be the same size as the first input, x. Recall that the convolution is performed by sliding the flipped kernel along the signal. Visualize sliding k along x in figure A.3. If we slide the kernel all the way to the leftmost part of the signal, then it slides off the edge where x is not defined, and the same thing happens on the right edge. Specifically, it slides off when we're trying

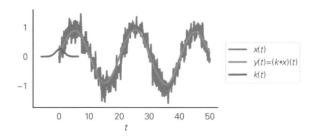

Figure A.3
Smoothing a noisy signal by convolving with a Gaussian kernel. A noisy sine wave, $x(t)$, is convolved with a Gaussian kernel, $k(t)$, and the results is a smoothed version of the signal, $y(t) = (k * x)(t)$. Code to produce this figure can be found in `ConvolutionExample.ipynb`.

to compute $y(t_0)$ for t_0 that is within one kernel-radius of each edge. The kernel-radius is `3*sigma` in this code. One solution to this problem is to slide the kernel only as far as it can go without sliding off the edge, but then we could not compute $y(t)$ at values of t that are close to the edge, so $y(t)$ would necessarily be shorter than $x(t)$ by two kernel radii. If we wanted to use this option, we'd use the option `'valid'` in place of `'same'`. The `'same'` option produces a `y` that is the same size as `x` by sliding the kernel past the edges of `x` and assuming that $x(t)$ is zero beyond its edges. This is called "zero padding" because it is equivalent to padding `x` with zeros on either side, then performing a valid convolution. Problems known as *boundary effects* can arise when $x(t)$ very far from zero at its edges. These boundary effects only affect the value of $y(t)$ within a kernel-radius of each edge. So long as our kernel-radius is much shorter than our signals, these boundary effects might not be a big deal.

- We multiply the output of `convolve` by `dt` in the last line. This is because `convolve` performs a discrete-time convolution, which is defined by sums instead of integrals. Multiplying by `dt` turns these sums into Riemann sums that approximate the integrals in equation (A.11).

Exercise A.3.1. Run `ConvolutionExample.ipynb` and try changing the signal, kernel, and other variables to get a feel for how convolutions work. For example, adding a large constant to `x` (moving it up) can help visualize boundary effects. Changing the kernel-radius will exaggerate these effects. Try changing the kernel to implement a causal filter. What do boundary effects look like for causal kernels? Why?

A.4   One-Dimensional Linear ODEs with Time-Dependent Forcing

We now consider a single-variable ODE driven by a time-dependent forcing term:

$$\tau \frac{dy}{dt} = -y + I(t)$$

(A.12)

$$y(0) = y_0$$

where $I(t)$ is some function of time. For simplicity, we assumed an initial condition at $t_0 = 0$ in equation (A.12) because solutions with initial conditions of the form $y(t_0) = y_0$ for $t_0 \neq 0$ are identical, but shifted in time.

If $I(t) = c$, then equation (A.12) is equivalent to equation (A.8), which produces exponential decay. Now, we are interested in solutions with time-dependent $I(t)$. The function, $I(t)$, is sometimes called a *forcing term*.

To get an intuition for solutions with time-dependent $I(t)$, first consider taking a step-function forcing term:

$$I(t) = \begin{cases} 0 & t < t_1 \\ c & t \geq t_1. \end{cases}$$

(A.13)

This function starts at 0 and then jumps to c at some time $t_1 > 0$.

Without solving the equation for this $I(t)$, we can already visualize what the solution should look like. Until time t_1, we are just solving equation (A.3), so $y(t)$ will decay exponentially toward zero from the initial condition, y_0. After time t_1, we are solving

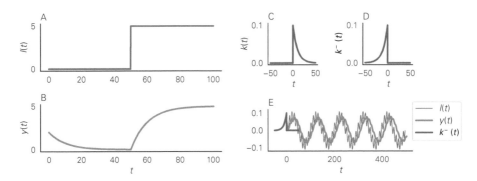

Figure A.4
Solutions to linear ODEs with time-dependent forcing. (A) A step function forcing term, $I(t)$, defined by equation (A.13) and (B) the resulting solution, $y(t)$, to equation (A.13) given by equations (A.14) and (A.15) with $c = 5$, $y(0) = 2$, and $\tau = 10$. (C) The kernel, $k(s)$, defined by equation (A.17) and the flipped kernel, $k^-(s) = k(-s)$. (D) The solution, $y(t)$ under an oscillating forcing term, $I(t)$. The solution is obtained by sliding k^- along $I(t)$ and computing a weighted average (compare to figures A.2 and A.3). Code to produce these plots can be found in `LinODE.ipynb`.

equation (A.5), so $y(t)$ will decay exponentially toward c. In other words,

$$y(t) = y_0 e^{-t/\tau}, \quad t < t_1 \tag{A.14}$$

and $y(t_1) = y_0 e^{-t_1/\tau}$, so

$$y(t) = c + (y_0 e^{-t_1/\tau} - c)e^{-(t-t_1)/\tau}, \quad t \geq t_1. \tag{A.15}$$

We are basically gluing together two solutions with constant $I(t)$. See figure A.4a,b for an example of this solution. Technically, the solution we found is not differentiable at $t = t_1$, so equation (A.12) is not satisfied at $t = t_1$, but it is still satisfied everywhere else, and this is the only solution that makes any sense, so we won't worry about it.

Of course, we can extend this idea to any piecewise constant function, $I(t)$. The solution to equation (A.12) will just decay exponentially toward the new value of $I(t)$ every time that $I(t)$ changes. But what about functions, $I(t)$, that change continuously in time? The same idea is valid: The solution, $y(t)$, is constantly trying to decay toward $I(t)$, but it can't catch up because $I(t)$ is always changing. This is easy to see if you approximate a continuous $I(t)$ with a piecewise constant function that is constant over very short time intervals, which is exactly what we do when we define functions on discretized time intervals in Python.

Specifically, the solution to equation (A.12) is given by

$$y(t) = y_0 e^{-t/\tau} + (k * I)(t) \tag{A.16}$$

where $*$ denotes convolution and

$$k(s) = \frac{1}{\tau} e^{-s/\tau} H(t) = \begin{cases} \frac{1}{\tau} e^{-s/\tau} & s \geq 0 \\ 0 & s < 0 \end{cases} \tag{A.17}$$

is an exponential kernel where $H(t)$ is the Heaviside step function ($H(s) = 1$ for $s \geq 0$ and $H(s) = 0$ for $s < 0$). You can verify that equation (A.16) satisfies equation (A.12) by direct substitution. See figure A.4c,d,e for a visualization of this solution.

The function, $k(s)$, is called the "Greens function" for the ODE in equation (A.12). Note that $k(s) = 0$ for $s < 0$, so this is a causal filter. In other words, $y(t)$ only depends on $I(s)$ for $s < t$. Also note that

$$\int_{-\infty}^{\infty} k(s)ds = 1$$

so $y(t)$ represents a running, weighted average of $I(t)$. Specifically, the value of $y(t)$ is given by the average values of $I(s)$ over the past ($s < t$), weighted by an exponential that decays with time constant, τ. More recent values of $I(s)$ are weighted more heavily and values of $I(s)$ further in the past are weighted less heavily. The parameter τ determines how quickly $y(t)$ "forgets" past values of $I(s)$. Roughly speaking, $y(t)$ is affected only by values of $I(s)$ in the interval $[t - 5\tau, 0]$ because the interval $[0, 5\tau]$ contains the almost all of the "mass" of $k(s)$–that is, because $\int_0^{5\tau} k(s)ds = 0.993 \approx 1$.

Exercise A.4.1. Consider the case where there is an additive constant in the ODE:

$$\tau \frac{dy}{dt} = -y + c + I(t)$$

$$y(0) = y_0.$$

Show that the solution is

$$y(t) = y_0 e^{-t/\tau} + c + (k * I)(t).$$

A.5 The Forward Euler Method

We now consider a simple method for numerically approximating solutions to ODEs that is useful when we cannot find closed-form solutions. We first consider ODEs in one dimension of the form

$$\frac{dx}{dt} = f(x, t) \tag{A.18}$$

$$x(t_0) = x_0.$$

Numerically approximating solutions to differential equations is a major subject in applied mathematics, but most research is devoted to solving PDEs because ODEs are much easier to solve numerically. The idea behind most numerical approximations is to first rewrite equation (A.18) as

$$dx = f(x, t)dt.$$

This equation does not have a precise mathematical meaning because dx/dt does not represent a literal fraction. However, the equation can be interpreted to mean that

$$dx = x(t + dt) - x(t) \approx f(x(t), t)dt$$

when dt is sufficiently small. This can in turn be written as

$$x(t + dt) \approx x(t) + f(x(t), t)dt. \tag{A.19}$$

This is known as an *Euler step*. If we want to be more mathematically precise, suppose that we know the value of $x(t)$ and we interpret $x(t) + f(x(t), t)dt$ as an approximation to the value of $x(t + dt)$. Then the error made by this approximation is decays to zero faster than dt decays to zero, specifically

$$\lim_{dt \to 0} \frac{\text{error}}{dt} = \frac{x(t + dt) - (x(t) + f(x(t), t)dt)}{dt} = 0. \tag{A.20}$$

How can we use equation (A.19) to approximate solutions to the ODE in equation (A.18)? We are implicitly assuming that we already know the initial condition, $x(t_0)$. Now we can plug in $t = t_0$ into equation (A.19) to obtain an approximation to $x(t_0 + dt)$:

$$x(t_0 + dt) \approx x(t_0) + f(x(t_0), t_0)dt = x_0 + f(x_0, t_0).$$

Now that we have an approximation to $x(t_0 + dt)$, we can take another Euler step to obtain an approximation to $x(t_0 + 2dt)$:

$$x(t_0 + 2dt) \approx x(t_0 + dt) + f(x(t_0 + dt), t_0 + dt)dt$$

where we would need to plug in our previous approximation for $x(t_0 + dt)$. We can repeat this procedure to obtain approximations to $x(t_0 + 3dt)$, $x(t_0 + 4dt)$ and so on. In summary, if we start with a discretized time vector starting at $t = t_0$, then we can approximate the solution to equation (A.18) at time points along this vector. The algorithm is arguably simpler written in Python:

```
# Initialize x
x=np.zeros_like(time)

# set initial condition
x[0]=x0

# Loop through time and perform Euler steps
for i in range(len(time)-1):
    x[i+1]=x[i]+f(x[i],time[i])*dt
```

or we can replace `f(x(i),time(i))` in the loop with a string of code representing the right side of our ODE. This procedure for approximating $x(t)$ is called the *forward Euler method* or sometimes, just *Euler's method*. It is the simplest method for numerically approximating solutions to ODEs. Note that we can still apply the forward Euler method if the right side of our ODE does not depend on t. We just omit t and write $f(x)$. Everything else works out just the same. A more complete example of applying the forward Euler method is given in `ForwardEuler.ipynb`, and the results are plotted in figure A.5.

If we replace equation (A.19) by an approximation that gives higher powers of dt in the denominator of equation (A.20), like dt^2, we get a higher-order method, whereas the forward Euler method is a first-order method. With higher-order methods, you can use a larger time step, dt, and still get a smaller error. When dt is larger, the for loop will be shorter, so the method will run faster. Higher-order methods require smoothness assumptions on $f(x, t)$, $x(t)$, or both so they can't be directly applied to integrate-and-fire (IF) neuron models due to the discontinuity of $V(t)$ at reset, or to equations with Dirac delta functions like our synapse models. Firing-rate models can benefit from higher-order methods, but we will stick with

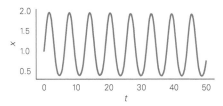

Figure A.5
Forward Euler method example. The numerical solution obtained by the forward Euler method with $f(x,t) = \sin(x)\cos(t)$, $dt = 0.01$, $t_0 = 0$, and $x_0 = 1$. The Code to produce this figure can be found in `ForwardEuler.ipynb`.

the forward Euler method in this textbook. See any good textbook on numerical analysis for more information on higher-order methods.

The forward Euler method is easily extended to ODEs in higher dimensions or "systems of ODEs." Consider the system of two ODEs:

$$\frac{dx}{dt} = f(x,y,t)$$

$$\frac{dy}{dt} = g(x,y,t)$$

$$x(t_0) = x_0, \ y(t_0) = y_0.$$

The forward Euler method for solving this system is essentially the same, but with two variables instead of one:

```
x=np.zeros_like(time)
y=np.zeros_like(time)
x[0]=x0
y[0]=y0
for i in range(len(time)-1):
    x[i+1]=x[i]+f(x[i],y[i],time[i])*dt
    y[i+1]=y[i]+g(x[i],y[i],time[i])*dt
```

Then we can use `plot(time,x)` and `plot(time,y)` to plot each solution or `plot(x,y)` to plot the trajectory of the solution in two dimensions. We can use the same approach to solve a system of any number of ODEs. If the equation is written in vector form:

$$\frac{d\boldsymbol{u}}{dt} = \boldsymbol{F}(\boldsymbol{u},t)$$

$$\boldsymbol{u}(0) = \boldsymbol{u}_0$$

where $\boldsymbol{u} = (x,y)$ is a vector, then we can use

```
u=np.zeros((2,len(time)))
u[:,0]=u0
for i in range(len(time)-1):
    u[:,i+1]=u[:,i]+F(u[:,i],time[i])*dt
```

Then we would use `plot(time,u(1,:))` and `plot(time,u(2,:))` or `plot
(u(1,:),u(2,:))` to plot the solution. The forward Euler method ODEs in more than
two dimensions is similar.

Exercise A.5.1. The Lorenz system is defined by

$$x' = 10(y - x)$$
$$y' = x(28 - z) - y$$
$$z' = xy - (8/3)z.$$

It was originally developed as a model of convection in the atmosphere. It's not an accurate
model, but it is widely studied for its mathematical properties. Use the forward Euler method
with a time step of $dt = 0.01$ to solve the Lorenz system over the time interval $[a, b] = [0, 20]$,
with initial conditions $x(0) = y(0) = z(0) = 1$. Plot the solution in three dimensions.

Exercise A.5.2. Modify the code in `ForwardEuler.ipynb` to solve an ODE for which
you know a closed-form solution (*e.g.*, an ODE from section A.2 or A.4), and compare the
true solution to the approximation obtained using the forward Euler method for different
values of dt.

A.6 Fixed Points, Stability, and Bifurcations in One-Dimensional ODEs

In this appendix, we consider ODEs where $f(x, t)$ does not depend on time:

$$\frac{dx}{dt} = f(x)$$

$$x(t_0) = x_0. \tag{A.21}$$

Equation (A.21) should be interpreted to mean that

$$x'(t) = f(x(t))$$

for all t in the domain under consideration. ODEs that can be written in the form of
equation (A.21), where the right side depends only on x, are called *autonomous ODEs*.
The exponential integrate-and-fire (EIF) model with time-constant input, $I(t) = I_0$, can be
written in this form by dividing both sides by τ and it is therefore an autonomous ODE.

Autonomous ODEs have the advantage that we can understand the behavior of solutions without needing to solve them explicitly, either in closed-form or numerically. This
approach to studying the behavior of solutions to autonomous ODEs without computing
solutions is sometimes called *dynamical systems theory*. The advantage to this approach
is that we can understand the behavior of solutions across a range of different parameter
values and initial conditions. Dynamical systems theory is a beautiful and fun area of mathematics to learn. This section and section A.8 cover some of the basic topics in dynamical
systems theory. For a more in-depth treatment, see Strogatz (2018), which is my favorite
mathematics book.

The most important concept for understanding the behavior of solutions to autonomous
ODEs is the notion of a fixed point. A *fixed point* to equation (A.21) is defined as a number,

x^*, for which

$$f(x^*) = 0.$$

Some texts use x_0 to denote a fixed point, but we are using x_0 to denote an initial condition, so we use x^* to denote a fixed point instead.

Now, consider what happens if we start at a fixed point ($x_0 = x^*$); that is, if our initial condition satisfies $f(x_0) = 0$. Let's first compute $x'(t_0)$ in that case:

$$x'(t_0) = f(x(t_0)) = f(x_0) = 0.$$

Hence, if $x(t)$ starts at a fixed point, its derivative starts at zero. Indeed, it is easy to verify that equation (A.21) is satisfied by the constant function

$$x(t) = x_0$$

when x_0 is a fixed point; that is, when $f(x_0) = 0$. Therefore, if an initial condition coincides with a fixed point, then the solution to the ODE is a constant function. In other words:

If x(t) *starts at a fixed point, it stays there.*

This very simple principle is our first step to understanding solutions to ODEs. To check your understanding, try Exercise A.6.1.

This discussion tells us how solutions behave when the initial condition is a fixed point, but this only gets us so far. What happens when initial conditions is not a fixed point? Considering the IVP given in equation (A.21), if the initial condition is not a fixed point, then $f(x_0) > 0$ or $f(x_0) < 0$. Let's first consider the case $f(x_0) > 0$. Then $x'(t_0) = f(x_0) > 0$. In other words, the solution, $x(t)$, is initially increasing. By the same argument, if $f(x_0) < 0$, then $x(t)$ is initially decreasing.

But we can go further than looking at the initial behavior of $x(t)$. Since the ODE is autonomous, we can use the sign of $f(x)$ to determine whether $x(t)$ is increasing or decreasing at *any* value of t. With this approach, we can understand the behavior of solutions using a *phase line*, which is a sign chart for $f(x)$ that indicates the sign of $x'(t)$ at any value of $x(t)$ (figure A.6a). In a phase line, fixed points are marked by dots on an x-axis. These dots break the axis into regions over which $f(x)$ is either positive or negative–that is, over which $x(t)$ is either increasing or decreasing. In each region, we draw an arrow pointing to the right if $f(x) > 0$ in that region and pointing to the left if $f(x) < 0$ in that region. These arrows indicate which direction $x(t)$ is moving when $x(t)$ is within that region. Figure A.6 illustrates a phase line for the ODE

$$x' = x^2 - 1.$$

There are two fixed points at $x^* = \pm 1$. Whenever $x(t)$ is in the region $-1 < x_0 < 1$, the solution is decreasing ($f(x(t)) = x'(t) < 0$). When $x(t)$ is in either of the two regions $x_0 < -1$ or $x_0 > 1$, the solution increases. Notably, solutions cannot cross a fixed point so all solutions are either constant (if they start at a fixed point), strictly increasing, or strictly decreasing.

We can use the phase line to understand the behavior of solutions to one-dimensional autonomous ODEs. For example, for the ODE drawn in figure A.6a, what is the behavior of the solution starting at $x(0) = 0$? The solution decreases (because the arrow in the region

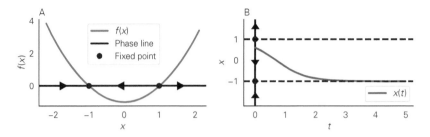

Figure A.6
Example of a phase line. (A) A phase line (black line with arrows) for the ODE $x' = f(x) = x^2 - 1$. The function $f(x)$ is plotted with a blue line, the fixed points are indicated by black dots, and the arrows indicate whether $x(t)$ increases (rightward facing) or decreases (leftward facing) on each side of the fixed points. (B) The phase line can be flipped onto the vertical axis to visualize solutions. Fixed points produce horizontal asymptotes (dashed lines) that bound solutions (red line). See PhaseLine.ipynb for the code to generate this figure without the arrows.

containing $x = 0$ in figure A.6a is pointing to the left) and it continues to decrease toward the fixed point at $x = -1$. It never quite reaches the fixed point, but $\lim_{t \to \infty} x(t) = -1$. What about the solution starting at $x(0) = 2$? It increases forever. All this information can be read from the phase line.

Figure A.6b shows how you can flip the phase line onto the vertical axis to visualize the plot of $x(t)$ versus t. Fixed points create horizontal asymptotes (dashed lines) and solutions (red lines) follow the arrow inside the region containing the initial condition.

Notice that both arrows surrounding the fixed point at $x^* = -1$ in figure A.6a are pointing toward the fixed point. As a result, solutions that start near that fixed point converge toward it. This motivates the idea of stable fixed points. A fixed point is called *asymptotically stable* if solutions that start sufficiently close to the fixed point converge to the fixed point (see the following discussion for an explanation of why we need to specify "asymptotically"). Definition A.1 states this more precisely.

Definition A.1. A fixed point, x^*, to equation (A.21) is called asymptotically stable if there is an $\epsilon > 0$ such that whenever $|x_0 - x^*| < \epsilon$, $\lim_{t \to \infty} x(t) = x^*$.

Now let's look more closely at why the fixed point at $x^* = -1$ is asymptotically stable. Since x^* is a fixed point, $f(x^*) = 0$. To the left of the fixed point, $f(x) > 0$, which is why we drew a right-facing arrow. Immediately to the right of the fixed point, $f(x) < 0$, so we drew a left-facing arrow. In other words, $f(x)$ changed from positive to negative at $x^* = -1$, and this is what caused the arrows to point toward $x^* = -1$. A smooth function that changes from positive to negative at x^* is necessarily decreasing at x^*; that is, its derivative is negative. This motivates theorem A.1.

Theorem A.1. Suppose that x^* is a fixed point of equation (A.21). If $f'(x^*) < 0$, then x^* is an asymptotically stable fixed point.

Now, notice that both arrows surrounding the fixed point at $x^* = 1$ are pointing away from the fixed point. Therefore, solutions that start near $x^* = 1$ go away from it. This motivates the idea of unstable fixed points. A fixed point is *asymptotically unstable* if it is not stable; that is, if solutions can start arbitrarily close to the fixed point without converging to it. Definition A.2 states this point more specifically.

Definition A.2. A fixed point, x^*, to equation (A.21) is called asymptotically unstable if for any $\epsilon > 0$, there is an x_0 such that $|x_0 - x^*| < \epsilon$ and $\lim_{t \to \infty} x(t) \neq x^*$.

The fixed point at $x^* = 1$ is asymptotically unstable because $f(x) < 0$ to the left of the fixed point and $f(x) > 0$ to the right of the fixed point, so the arrows point away from $x^* = 1$. Mirroring the discussion above, this motivates theorem A.2.

Theorem A.2. Suppose that x^* is a fixed point of equation (A.21). If $f'(x^*) > 0$, then x^* is an asymptotically unstable fixed point.

The reason that we needed to specify "asymptotically" in the definitions of stable/unstable given here is that there are some borderline cases in which a fixed point is stable in some senses, but not others. Notably, for one-dimensional autonomous ODEs like equation (A.21) with smooth $f(x)$, these occur only when $f'(x^*) = 0$. To see what can happen when $f'(x^*) = 0$, first consider the fixed point $x^* = 0$ for $f(x) = x^2$. By the definitions here, this fixed point is asymptotically unstable, but some texts classify it as "semi-stable" since $x(t) \to x^*$ for initial conditions on one side of the fixed point, but not the other. Indeed, "asymptotically unstable" is sometimes defined in a way that excludes semi-stable fixed points. Regardless of how this fixed point is classified, the behavior of solutions is still easily understood by drawing a phase line. Note that $f'(x^*) = 0$ does not imply that a fixed point is semi-stable. Consider, for example, $f(x) = x^3$.

Another example where stability is less clear is given by the trivial ODE defined by a constant zero function, $f(x) = 0$. Every x is a fixed point, and the solution to equation (A.21) is constant, $x(t) = x_0$, for any initial condition $x(0) = x_0$. Solutions that start near a fixed point (but not at the fixed point) do not converge to that fixed point, but also do not travel away from it–that is, they stay near it. Some texts would classify these fixed points as "stable" but not "asymptotically stable," and some texts use the phrase "neutrally stable" for such fixed points.

The previous two paragraphs seem to cloud the waters around stability, but the core of the idea remains simple in almost all cases: Whenever $f'(x^*) < 0$ or $f'(x^*) > 0$, a fixed point is stable/unstable in every sense. When $f'(x^*) = 0$, stability can be trickier to classify, but a phase line can still help understand the behavior of solutions. Generally, *drawing a phase line is an easier and more informative way to understand properties of a fixed point than computing $f'(x^*)$.* To simplify the terminology, we will drop the "asymptotically" label and simply refer to fixed points with $f'(x^*) < 0$ or $f'(x^*) > 0$ as *stable* or *unstable* when there is no ambiguity.

One of the most useful things about the dynamical systems approach to ODEs described in this section is that we can understand the behavior of solutions to ODEs without even solving them. This is especially useful because it lets us study how solutions depend on parameter values. For example, the existence and stability of fixed points can depend on the parameters that define an ODE. A parameter value at which the number or stability of fixed points changes is called a *bifurcation*. The study of bifurcations is a central theme in dynamical systems. The most common type of bifurcation in one-dimensional ODEs is a *saddle-node bifurcation* in which two fixed points collide and disappear as a parameter is changed. Exercise A.6.4 demonstrates a saddle-node bifurcation.

Exercise A.6.1. Consider the IVP

$$x' = x^2 - 1$$

$$x(0) = 1.$$

Find the solution, $x(t)$, in closed form (don't overthink it; this step should be very easy if you think about the previous discussion). Use the forward Euler method to compute a numerical solution with $dt = 0.01$ and $T = 5$. Compare the numerical solution to the closed-form solution. Now, compute the first couple of Euler steps by hand and think about why the forward Euler method gives the solution that it does.

Exercise A.6.2. Consider the IVP

$$x' = x^2 - 1$$

$$x(0) = -2.$$

What is the behavior of the solution $x(t)$? What is the sign of $x'(3)$? What is $\lim_{t \to \infty} x(t)$? Reproduce figure A.6b with $x(0) = -2$ to check your answer.

Exercise A.6.3. Consider the ODE

$$x' = -(x+1)(x-1)(x-2).$$

Draw a phase line, find all fixed points, and classify their stability. Solve the equation numerically using the forward Euler method with different initial conditions to verify your results.

Exercise A.6.4. Consider the ODE

$$x' = x^2 + a$$

that depends on the parameter, a. Draw a phase line and classify all fixed points and their stability for the three cases: $a > 0$, $a = 0$, and $a < 0$. There is a saddle-node bifurcation at $a = 0$.

A.7 Dirac Delta Functions

When modeling spike trains and when modeling ODEs with time-dependent forcing, we often want to model a very fast pulse. We can define a pulse of width Δt at time $t = 0$ as

$$I(t) = \begin{cases} \frac{1}{\Delta t} & t \in [-\Delta t/2, \Delta t/2] \\ 0 & \text{otherwise.} \end{cases}$$

This is just a rectangle of width Δt and height $1/\Delta t$ centered on $t = 0$. Note that whenever $a > \Delta t/2$,

$$\int_{-a}^{a} I(t)dt = 1.$$

That is, the area under the rectangle is 1. A fast pulse is modeled by taking small values of Δt. As a mathematical abstraction, it is often useful to consider an infinitely fast pulse by taking $\Delta t \to 0$. However, note that $I(0) \to \infty$ in this limit, so $I(t)$ does not converge to a function in the usual sense. Dirac delta functions give us a way to work with infinitely fast pulses and treat them like functions.

The *Dirac delta function*, $\delta(t)$, is a "function" defined by $\delta(t) = 0$ for $t \neq 0$ and

$$\int_{-a}^{a} \delta(t)dt = 1$$

for any $a > 0$. We will use the term *delta function* as shorthand for the Dirac delta function.

We put "function" in quotation marks here because the Dirac delta function is not actually a function in the strict sense of the word. Any real function that satisfies $\delta(t) = 0$ for $t \neq 0$ would also satisfy $\int_{a}^{b} \delta(t)dt = 0$ for all $a, b \in \mathbb{R}$. Intuitively, we can think of a Dirac delta function as an infinitely narrow and infinitely tall pulse, so $\delta(0) = \infty$. There is a mathematically precise way to define Dirac delta functions as a "distribution," "generalized function," or "measure." Under these definitions, the delta function can be evaluated only inside an integral, so we never need to ask about the value of $\delta(0)$.

The delta function can also be interpreted as a Gaussian probability density with mean zero and standard deviation zero:

$$\delta(t) = \lim_{\sigma \to 0} \frac{1}{\sigma\sqrt{2\pi}} e^{-t^2/(2\sigma^2)}$$

and this interpretation also represents the limit of an infinitely narrow pulse at $t = 0$.

We often want to consider a pulse centered at some value of $t \neq 0$. To do this, we can just translate $\delta(t)$ and define

$$\delta_{t_1}(t) = \delta(t - t_1)$$

which has the property

$$\int_{a}^{b} \delta_{t_1}(t)dt = \int_{a}^{b} \delta(t - t_1)dt = \begin{cases} 1 & a < t_1 < b \\ 0 & \text{otherwise} \end{cases}$$

The translated pulse should be interpreted as an infinitely narrow pulse centered at $t = t_1$. Dirac delta functions have the following important property:

$$\int_{-a}^{a} x(t)\delta(t)dt = x(0)$$

for $a > 0$ and, more generally,

$$\int_{a}^{b} x(t)\delta(t - t_1)dt = \int_{a}^{b} x(t)\delta_{t_1}(t)dt = x(t_1)$$

when $a < t_1 < b$. Indeed, in a more mathematically rigorous setting, this is essentially the definition of the delta function. As a consequence, delta functions are identities under convolution:

$$(x * \delta)(t) = x(t)$$

and convolution with a translated delta functions implements a translation:

$$(x * \delta_{t_1})(t) = x(t - t_1). \tag{A.22}$$

Now, consider what happens when a Dirac delta function is the forcing term in a one-dimensional linear ODE:

$$\tau \frac{dx}{dt} = -x + \delta(t - t_1)$$

$$x(0) = 0 \tag{A.23}$$

and let's assume that $t_1 > 0$. The first way to approach this problem is to use the solution derived in section A.4. Specifically, from equation (A.16), we have

$$x(t) = (k * \delta_{t_1})(t)$$

where $\delta_{t_1}(t) = \delta(t - t_1)$ is the forcing term and $k(s) = \frac{1}{\tau}e^{-s/\tau}H(s)$ is an exponential kernel with $H(t)$ being the Heaviside step function. From equation (A.22), therefore, we have

$$x(t) = \frac{1}{\tau}e^{-(t-t_1)/\tau}H(t - t_1) = \begin{cases} \frac{1}{\tau}e^{-(t-t_1)/\tau} & t \geq t_1 \\ 0 & t < t_1. \end{cases} \qquad (A.24)$$

In other words, $x(t) = 0$ for $t < t_1$, and then it jumps up to $1/\tau$ at time $t = t_1$ and then decays back toward zero for $t > t_1$. Put more simply, $x(t)$ is just the exponential kernel, $k(t)$, shifted in time by t_1.

So far, we have considered the mathematical definition of a Dirac delta function, but how should we represent it numerically in code? If we are using discretized time with a step size of dt, we can represent a delta function by placing $1/dt$ in the associated time bin. For example, to represent the signal $I(t) = \delta(t - t_1)$, we would do

```
time=np.arange(0,T,dt)
I=np.zeros_like(time)
I[int(t1/dt)]=1/dt
```

This code assumes that $0 \leq t_1 < T$. Figure A.7a shows a numerical representation of a delta function at $t_1 = 20$ with $dt = 0.1$. This way of representing Dirac delta functions interacts nicely with Riemann integration, discrete convolutions, and the forward Euler method. For example, if we use the Riemann integral to approximate $\int_0^T I(t)dt$ by computing

```
integral0T=sum(I)*dt
```

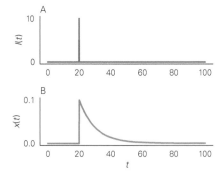

Figure A.7
Numerical representation of a Dirac delta function driving a linear ODE. (A) A numerical representation of $I(t) = \delta(t - t_1)$ with $t_1 = 20$ using a bin size of $dt = 0.1$. (B) Numerical solution of equation (A.23) obtained using the $I(t)$ from A with $\tau = 10$. The code to produce this figure can be found in `DiracDeltaFunctions.ipynb`.

then we will get the correct result because the `*dt` in this line of code cancels the `1/dt` in the associated bin in I to give 1. To see how numerical representations of Dirac delta functions interact with discrete convolutions and the forward Euler method, let's consider the numerical solution of equation (A.23). The forward Euler method would look like the following:

```
x[0]=0
for i in range(len(time)-1):
    x[i+1]=x[i]+dt*(-x[i]+I[i])/tau
```

In this loop, `x[i+1]` will remain zero until we reach the time bin `i==np.int(t1/dt)` corresponding to time t_1. After that time bin, x will jump up to `x[i+1]=dt*(1/dt)/tau`, which is equal to `1/tau`. After that, x will decay back to zero. This is exactly the behavior of the true solution in equation (A.24). Similarly, we can compute the solution from equation (A.23) using a discrete convolution:

```
s=np.arange(-5*tau,5*tau,dt)
k=(1/tau)*np.exp(-s/tau)*(s>=0)
x=np.convolve(I,k,'same')*dt
```

which gives similar results. Figure A.7b shows a numerical solution to equation (A.23). The file `DiracDeltaFunctions.ipynb` contains code that produces figure A.7b using the forward Euler method and a numerical convolution.

A.8 Fixed Points, Stability, and Bifurcations in Systems of ODEs

In section A.6, we looked at fixed points and stability for autonomous ODEs in one dimension. We now extend some of those results to systems of ODEs, which can be viewed as ODEs in higher dimensions (i.e., $n > 1$). Recall that an autonomous system of ODEs can be written in the form

$$\frac{d\boldsymbol{u}}{dt} = \boldsymbol{F}(\boldsymbol{u}) \tag{A.25}$$

where $\boldsymbol{u}: \mathbb{R} \to \mathbb{R}^n$ and $\boldsymbol{F}: \mathbb{R}^n \to \mathbb{R}^n$. Given a system of ODEs and an initial condition, $\boldsymbol{u}(0) = \boldsymbol{u}^0$, the system has a unique solution satisfying the initial condition (under some assumptions on $\boldsymbol{F}$). A system can also be written as a list of one-dimensional equations; for example, if $n = 2$, then

$$\frac{dx}{dt} = f(x, y)$$
$$\frac{dy}{dt} = g(x, y) \tag{A.26}$$

where $x, y: \mathbb{R} \to \mathbb{R}$, and the two conventions are related by defining

$$\boldsymbol{u}(t) = \left[\begin{array}{c} x(t) \\ y(t) \end{array} \right].$$

As for one-dimensional ODEs, a *fixed point* for the system in equation (A.25) is again defined as a value, $u^* \in \mathbb{R}^n$, satisfying

$$F(u^*) = 0.$$

Fixed points again satisfy the following property:

If $u(t)$ starts at a fixed point, it stays there.

In other words, if $u^0 = u^*$, where $F(u^*) = 0$, then $u(t) = u^*$ for all t. In addition, if a solution does not start at a fixed point, then it cannot ever equal 1. In other words, if $F(u^0) \neq 0$, then $F(u(t)) \neq 0$ for all t.

The definitions of stable and unstable fixed points are also the same for systems as for one-dimensional ODEs: A fixed point is stable if initial conditions sufficiently close to the fixed point converge to it, and it is unstable otherwise.

However, determining whether a fixed point is stable or unstable is more complicated for systems of ODEs. We will begin by considering *linear systems of ODEs*, which are systems for which $F(u) = Au$ is a linear function, so

$$\frac{du}{dt} = Au$$

$$u(0) = u^0$$

(A.27)

for some $n \times n$ matrix, A. Linear systems always have a fixed point at the zero vector:

$$u^* = 0.$$

If A is nonsingular (*i.e.*, invertible), then this is the only fixed point. The stability of the fixed point at zero is determined by the eigenvalues of A. Recall that eigenvalues are numbers, λ, for which there exists a vector, v, that satisfies

$$Av = \lambda v.$$

The vector, v, is called the eigenvector associated with the eigenvalue λ. Eigenvalues and eigenvectors can be real, imaginary, or complex. Complex eigenvalues always come in conjugate pairs, $\lambda = a \pm bi$. In general, an $n \times n$ matrix has n eigenvalues. To compute eigenvalues, note that $Av = \lambda v$ implies that $[A - \lambda Id]v = 0$ $A - \lambda Id$ is a singular matrix (where Id is the $n \times n$ identity matrix). Hence, we only need to solve $\det(A - \lambda Id) = 0$ for λ. Recall that, for a 2×2 matrix, the determinant is

$$\det\left(\begin{bmatrix} a & b \\ c & d \end{bmatrix}\right) = ad - bc.$$

In NumPy, eigenvalues and eigenvectors can be found by

```
lam,v=np.linalg.eig(A)
```

which returns a vector of all eigenvalues in `lam` and a matrix of eigenvectors in v.

The stability of linear systems is determined by the sign of the real part of the eigenvalues of A, as explained in theorem A.3,

Theorem A.3. If *all* eigenvalues of A have a *negative real part*, then the fixed point at zero is *stable* for equation (A.27). If *at least one* eigenvalue of A has a *positive real part*, then the fixed point at zero is *unstable* for equation (A.27).

We will not consider systems for which A has eigenvalues with real part equal to zero.

Figure A.8 shows numerical solutions of linear systems in $n = 2$ dimensions for various As with different eigenvalues. Two systems are stable and three are unstable. Note that the qualitative appearance of the solutions are quite different: In some cases, the solutions spiral in or out, but in other cases they do not. These properties are also determined by the eigenvalues. In particular, consider equation (A.27) in $n = 2$ dimensions. If the eigenvalues of A are complex, then solutions draw spirals in the plane.

When the eigenvalues have a negative real part, this is called a *stable spiral* or *spiral sink* and when they have a positive real part, this is called an *unstable spiral* or *spiral source*. Spirals in the $u(t)$ plane translate to oscillations of the individual components, $u_1(t)$ and $u_2(t)$, as shown in figure A.9.

When all eigenvalues are real and negative, solutions just decay exponentially to zero, and the system is called a *stable node* or *nodal sink*. When all eigenvalues are real and positive, solutions grow exponentially, and the system is called an *unstable node* or *nodal source*. When all eigenvalues are real but have different signs, solutions decay in some directions, but almost all solutions eventually grow exponentially. This is called a *saddle*. All of these solution types can be found in figures A.8 and A.9.

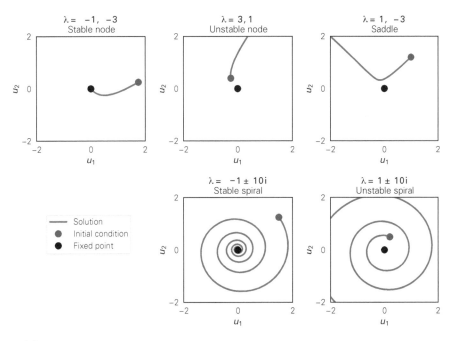

Figure A.8
Solutions of linear systems of ODEs with different eigenvalue patterns. The system in equation (A.27) solved using the forward Euler method for five different matrices, A. Eigenvalues, λ appear in the titles. The code to produce this figure can be found in `LinearSystemsOfODEs.ipynb`.

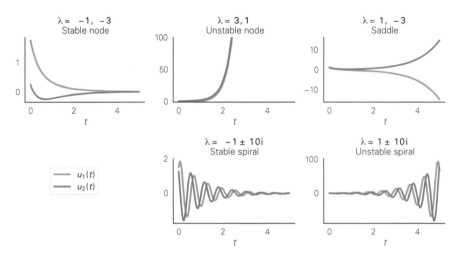

Figure A.9
Solutions of linear systems of ODEs with different eigenvalue patterns. It is the same as in figure A.8, but $u_1(t)$ and $u_2(t)$ are plotted as functions of t. The code to produce this figure can be found in `LinearSystems OfODEs.ipynb`.

Two-dimensional systems can be classified without even computing the eigenvalues. First, note that the determinant of a matrix is the product of its eigenvalues and the trace of a matrix is the sum of the eigenvalues. For $n = 2$,

$$\det(A) = \lambda_1\lambda_2, \quad \text{Tr}(A) = \lambda_1 + \lambda_2. \tag{A.28}$$

Recall that the trace is defined as the sum of the diagonal entries:

$$\text{Tr}\left(\begin{bmatrix} a & b \\ c & d \end{bmatrix}\right) = a + d$$

and the determinant of a 2×2 matrix is given by

$$\det\left(\begin{bmatrix} a & b \\ c & d \end{bmatrix}\right) = ad - bc.$$

From equation (A.28), it can be shown that a two-dimensional system has eigenvalues with a negative real part (and is therefore stable) if and only if

$$\det(A) > 0 \text{ and } \text{Tr}(A) < 0.$$

We can also use equation (A.28) to classify the type of system. If the determinant is negative, then the eigenvalues must be real (since $(a + bi)(a - bi) = a^2 + b^2 \geq 0$) and must have opposite signs, so the system must be a saddle. Distinguishing between nodes and spirals is a little bit trickier. From equation (A.28), it can be shown that

$$\lambda_{1,2} = \frac{T \pm \sqrt{T^2 - 4D}}{2} \tag{A.29}$$

where $T = \text{Tr}(A)$ and $D = \det(A)$. From this, we can show that $T^2 > 4D$ produces real eigenvalues (a node), and $T^2 < 4D$ means complex eigenvalues (a spiral). Putting this together, we can make a picture of the *trace-determinant plane*, in which the five types

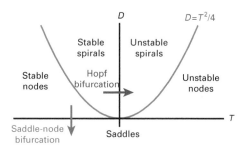

Figure A.10
The trace-determinant plane. The plane of all values of the trace (T) and determinant (D) of the matrix A from equation (A.27). The plane is split into five regions by the lines $T = 0$ and $D = 0$, along with the curve $D = T^2/4$. Each of these regions represents a different type of solution to equation (A.27). A Hopf bifurcation occurs during a transition between a stable and unstable spiral; that is, when T changes sign with $D > 0$.

of solutions discussed here are represented by five regions on the plane of all T and D values (figure A.10). Note that equation (A.29) can also be used to compute the eigenvalues directly, but it is valid only in $n = 2$ dimensions.

We have so far only considered stability for *linear* systems of ODEs. It turns out that we can use what we learned for linear systems to determine the stability of nonlinear systems.

Recall that for one-dimensional ODEs, stability is determined by the sign of $f'(x)$ at the fixed point. A similar result holds for systems of ODEs, but we need to generalize the notion of a derivative to a vector function $\boldsymbol{F} : \mathbb{R}^n \rightarrow \mathbb{R}^n$. The derivative of such a function is a matrix called a *Jacobian matrix*. Specifically, the Jacobian matrix of $\boldsymbol{F}$ at a point $\boldsymbol{u}^*$ is an $n \times n$ matrix, J, with entries defined by

$$J_{jk} = \frac{\partial F_j}{\partial u_k}\bigg|_{\boldsymbol{u} = \boldsymbol{u}^*}.$$

In other words, the j, kth entry of the Jacobian is the derivative of the j entry of $\boldsymbol{F}$ with respect to the kth entry of its input, evaluated at the fixed point. This equation is easier to understand when we write it out for a system written in the form of equation (A.26). For this system, the Jacobian is given by

$$J = \begin{bmatrix} \frac{\partial f}{\partial x} & \frac{\partial f}{\partial y} \\ \frac{\partial g}{\partial x} & \frac{\partial g}{\partial y} \end{bmatrix}$$

where the derivatives are evaluated at the fixed point in question.

Jacobian matrices are derivatives in the sense that they represent the best linear approximation to $\boldsymbol{F}$ at the fixed point. Specifically, Taylor's theorem for vector functions tells us that for the values of $\boldsymbol{u}$ near $\boldsymbol{u}^*$, we have

$$\boldsymbol{F}(\boldsymbol{u}) = \boldsymbol{F}(\boldsymbol{u}^*) + J[\boldsymbol{u} - \boldsymbol{u}^*] + \mathcal{O}\left(\|\boldsymbol{u} - \boldsymbol{u}^*\|^2\right)$$

where $\|\cdot\|$ is the Euclidean norm and the $\mathcal{O}(\cdot)$ term represents errors that go to zero quadratically as $\boldsymbol{u} \rightarrow \boldsymbol{u}^*$. If $\boldsymbol{u}^*$ is a fixed point, then the first term is $\boldsymbol{F}(\boldsymbol{u}^*) = 0$, and we have

$$\frac{d\boldsymbol{u}}{dt} = J[\boldsymbol{u} - \boldsymbol{u}^*] + \mathcal{O}\left(\|\boldsymbol{u} - \boldsymbol{u}^*\|^2\right)$$

for u near u^*. Defining a new variable $v(t) = u(t) - u^*$, we see that $v'(t) = u'(t) = F(u)$, so

$$\frac{dv}{dt} = F(u) = Jv + \mathcal{O}(\|v\|^2)$$

for $v(t)$ near its fixed point at $v^* = 0$. We may conclude the following general principle:

The solution of a nonlinear system like equation (A.25) or equation (A.26) near its fixed point behaves similarly to the solution to linear the system

$$\frac{dv}{dt} = Jv$$

near zero.

We can therefore use our understanding of linear systems to understand the behavior of nonlinear systems near their fixed points. In particular, this means that the eigenvalues of the Jacobian tell us about the stability of the fixed point, as stated in theorem A.4.

Theorem A.4. Consider the system given by equation (A.25) or equation (A.26), and let J be the Jacobian matrix computed at a fixed point. If *all* eigenvalues of J have a *negative real part*, then the fixed point is *stable*. If *at least one* eigenvalue of J has a *positive real part*, then the fixed point is *unstable*.

Moreover, if the Jacobian has complex eigenvalues, then solutions will tend to draw spirals, just as in the associated linear system.

For one-dimensional ODEs, we looked at saddle-node bifurcations in which changing a parameter of the ODE changed the stability of a fixed point. Systems of ODEs can exhibit saddle node bifurcations as well when the sign of an eigenvalue changes sign to switch between a stable node and a saddle (hence the name). Looking a the trace-determinant plane (figure A.10), we can see that this happens when D changes sign with $T > 0$.

Systems of ODEs can produce a different kind of bifurcation that cannot be produced by one-dimensional ODEs. When the stability of a spiral changes, it is called a *Hopf bifurcation*. Looking a the trace-determinant plane (figure A.10), we can see that this happens when T changes sign with $D > 0$. When a stable spiral becomes unstable, it usually gives rise to a *stable limit cycle*, which is a periodic solution to which nearby solutions converge. The periodicity of limit cycles implies that they generate oscillations.

Exercise A.8.1. Create an arbitrary 2×2 matrix on your own. Compute the eigenvalues by hand, and determine the stability and classification of the system. Now, use the forward Euler method to solve the system numerically, and compare the solution to what you expected from your classification. Try this a couple of times.

Exercise A.8.2. Repeat exercise A.8.1, but use the trace and determinant to determine the stability and classification without computing eigenvalues.

Exercise A.8.3. Consider the nonlinear system

$$\frac{\partial x}{\partial t} = -(1 - y^2)x - y$$

$$\frac{\partial y}{\partial t} = x.$$

Find the unique fixed point, find its stability, and classify it as node, spiral, or saddle. Then use the forward Euler method to simulate the system. Try different initial conditions and adjust the durations of the solution (T) to make sure that you can see the asymptotic behavior.

Exercise A.8.4.   Consider the nonlinear system

$$\frac{\partial x}{\partial t} = -c(1 - y^2)x - y$$

$$\frac{\partial y}{\partial t} = x$$

where μ is a parameter. Note that when $c = 1$, this is the same system considered in exercise A.8.3. Determine a value of c at which a Hopf bifurcation occurs. Then use the forward Euler method to simulate the system with different values of c on either side of the bifurcation. Plot the solution in the x-y plane (like the plots in figure A.8), and also plot the solutions as functions of time (like the plots in figure A.9) to see the oscillations.

Appendix B: Additional Models and Concepts

B.1 Ion Channel Currents and the HH Model

The first model of the ion channel dynamics underlying action potential generation was developed in the 1950s by Alan Hodgkin and Andrew Huxley in the 1950s (Hodgkin and Huxley 1952). Hodgkin and Huxley began their work by performing experiments on the squid giant axon (not to be confused with a giant squid axon) to study its electrical properties. They ultimately derived a mathematical description of an action potential and won the Nobel Prize in Physiology or Medicine (along with John Eccles) in 1963 for this work.

The model that they constructed is now referred to as the *Hodgkin-Huxley (HH) model*. For modeling other types of neurons in other species, parameters can be changed and other channel types can be added. Models that have the same basic structure as the HH model, but with different parameters or added ion channels, are called *HH-style models* and are widely used today. Most neuron models used today can be viewed as either simplifications or extensions of the HH model. This section provides a brief review of the original HH model. A more in-depth treatment can be found in many textbooks (e.g., Dayan and Abbott 2001; Gerstner et al. 2014) and websites.

There are two general categories of ion channels. *Ungated (passive) channels* are always open, so they allow ions to pass through all the time. *Gated (active) channels* open and close.

There are two factors that contribute to the flow of ions across an open channel: the concentration gradient and the electrical gradient. The effect of the concentration gradient is simple: When the concentration of a certain type of ion is greater inside the cell than outside the cell, then the concentration gradient tries to push this type of ion out. When the concentration is higher outside the cell, the concentration gradient tries to pull them in. In other words, the concentration gradient tries to equilibrate the inside and outside concentrations of each type of ion.

However, open channels do not simply cause inside and outside ion concentrations to equilibrate because the electrical potential also plays a role. As mentioned before, the membrane potential is usually negative. This negative potential tries to pull positive ions inside the cell and push negative ions out. The strength of this force depends on the neuron's membrane potential: A more negative membrane potential will push and pull the ions more strongly.

An ion's *reversal potential* is the membrane potential at which the effects of the concentration gradient and the electrical gradient cancel out. This is also sometimes called the *Nernst potential*. The reversal potential for ion type a is typically denoted as E_a (e.g., E_{Na} for sodium channels). When the membrane potential is at the reversal potential ($V = E_a$), the channel produces no current because the effects of the electrical and diffusive forces cancel each other out. In other words, the net flow of ions is zero (i.e., just as many flow in as out).

What about when $V \neq E_a$? In this case, there will be a flow of ions–that is, a current. Ion channels act as resistors; they allow ions to pass through, but with some resistance. The equation that describes the currents across an ion channel is

$$I_a = -g_a(V - E_a)$$

where g_a is the *conductance* of the channel, which measures how easily ions pass through ($g_a = 1/R_a$, where R_a is the resistance). This is similar to the leak current from section 1.1 in chapter 1, and it has the same interpretation: When $V > E_a$, there is a negative current that tries to pull V back down toward E_a. When $V < E_a$, there is a positive current that tries to pull V up toward E_a. So the membrane potential is pulled toward E_a by the ion channel. Therefore, under normal conditions, ion channels with reversal potentials above rest ($E_a > -72\,\text{mV}$) are depolarizing, while those with reversal potentials well below rest ($E_a < -72\,\text{mV}$) are hyperpolarizing. The currents induced by several parallel ion channels combine additively, so the total current can be modeled by a sum:

$$I_{total} = -\sum_a g_a(V - E_a).$$

The effect of individual channels of the same type can be lumped together, so we can get away with one term in the sum for each *type* of ion channel, a, instead of each individual channel.

For gated ion channels, $g_a(t)$ changes based on the number of ion channels that are open. Since a single neuron contains a large number of channels for each channel type, the *number* of open channels is approximately proportional to the *probability* that each channel is open. We can therefore write

$$g_a(t) = \overline{g}_a p_a(t)$$

where $p_a(t)$ is the probability that each ion channel is open and $\overline{g}_a$ would be the conductance if all channels were open. The open probability of gated ion channels can depend on many things, such as the membrane potential and the concentration of different ions, neuromodulators, or neurotransmitters.

The opening and closing of ion channels depends on complicated biophysical processes. A common approximation is to assume that the channel being open requires the opening of one or more gates. The opening of a gate requires multiple identical, independent processes to all be in an active state. We will refer to these processes as *subgates*. Consider an ion channel with one gate and k subgates. If $n(t)$ is the probability that each subgate is active, then the open probability of the channel is

$$p_a(t) = n^k(t).$$

The variable $n(t)$ is called an *activation variable* or a *gating variable*. In reality, the opening and closing of ion channels are more complicated, and parameters like k are typically fit to data instead of being derived from biophysical properties of the channel. For *voltage-gated ion channels*, n depends on the membrane potential, V. But note that V also changes in response to changes in n since open channels induce a current. This feedback loop between gating variables and membrane potentials makes it difficult to study the effects of channels, but it also imparts neurons with most of their interesting and useful properties. To start with, we will get around this bidirectional dependence by assuming the neurons are clamped to a fixed voltage.

In actual recordings, the voltage can be held fixed using a *voltage clamp*, which is a recording protocol in which the membrane potential (voltage) is held at specified fixed value by an electrode. This allows currents to be measured at fixed voltage. Voltage clamp is used to study properties of ion channels.

The first voltage-dependent ion channel that we will study here is the *voltage-dependent potassium (K_v) channel*. For the K_v channel, there is one gate with $k = 4$ subgates, so

$$I_K = -\overline{g}_K n^4 (V - E_K)$$

where $p_K(t) = n^4(t)$ is the probability that a K_v channel is open, or equivalently, the proportion of K_v channels that are open. Hodgkin and Huxley fit this model to experiments and found $\overline{g}_K = 36$ mS/cm^2, $E_K = -77$ mV (mS = millisiemens, a measure of conductance).

K_v is a voltage-dependent channel (hence the v subscript), so we also need to model the dependence of the gating variable, $n(t)$, on V. Recall that we are currently treating V as a constant since we are assuming that the cell is in a voltage clamp. The subgates open and close stochastically, and the rate at which they open and close depends on the voltage. Define $\alpha_n(V)$ to be the rate at which the closed subgates open and $\beta_n(V)$ to be the rate at which the open subgates close when the membrane potential is at V. This can be used to write a differential equation of the form

$$\frac{dn}{dt} = \alpha_n(V)(1 - n) - \beta_n(V)n. \tag{B.1}$$

The first term on the right side of the equation represents the rate at which newly opened subgates are added (by closed subgates becoming open). The second term on the right side quantifies the rate at newly closed subgates are added (by open subgates becoming closed).

While equation (B.1) is easy to interpret biophysically, it can be rewritten in a way that is easier to interpret dynamically:

$$\tau_n(V)\frac{dn}{dt} = n_\infty(V) - n$$

where

$$\tau_n(V) = \frac{1}{\alpha_n(V) + \beta_n(V)}$$

and

$$n_\infty(V) = \frac{\alpha_n(V)}{\alpha_n(V) + \beta_n(V)}.$$

When V is clamped (*i.e.*, constant), this is just a one-dimensional linear ordinary differential equation (ODE) for exponential decay ($\tau_n dn/dt = -n + n_\infty$), which has the following solution (see appendix A.2):

$$n(t) = (n_0 - n_\infty)e^{-t/\tau_n} + n_\infty.$$

In other words, when V is clamped, $n(t)$ decays exponentially to $n_\infty(V)$ with time constant $\tau_n(V)$. If $\tau_n(V)$ is large, it converges slowly. If $\tau_n(V)$ is small, it converges quickly.

The dependence of α_n and β_n on V is related to complicated molecular and cellular processes. Instead of modeling them explicitly, Hodgkin and Huxley empirically fit the dependence to experimental observations to get equations for α_n and β_n in terms of V:

$$\alpha_n(V) = \frac{0.01(V + 55)}{1 - e^{-0.1(V+55)}}$$

$$\beta_n(V) = 0.125e^{-0.0125(V+65)}$$

which have the unit $1/\text{ms} = \text{kHz}$ and V is measured in millivolts.

The other type of ion channel responsible for action potentials is the voltage-dependent sodium (Na_v) channel. Voltage-dependent sodium channels are more complicated than potassium channels because they produce a *transient conductance*, meaning that as V increases, they can switch from closed to open, and then back to closed again. This is because they essentially have two types of gates. When V increases, one opens quickly while the other closes slowly. To distinguish between these two gates, it is common to say "open" and "closed" for the fast gate and "active" and "inactive" for the slow gate.

The fast gate is represented by the gating variable m and has $k = 3$ subgates. The slow gate is represented with an h and has $k = 1$ subgates. This gives a current of the form

$$I_{Na} = -\bar{g}_{Na}m^3 h(V - E_{Na})$$

where $\bar{g}_{Na} = 120$ mS/cm^2, $E_{Na} = 50$ mV, m is the open probability, and h is the active probability. As before, the gating variables obey rate equations of the form

$$\frac{dm}{dt} = (1 - m)\alpha_m - m\beta_m$$

$$\frac{dh}{dt} = (1 - h)\alpha_h - h\beta_h$$

where

$$\alpha_m = \frac{0.1(V + 40)}{1 - e^{-0.1(V+40)}}$$

$$\beta_m = 4e^{-0.0556(V+65)}$$

$$\alpha_h = 0.07e^{-0.05(V+65)}$$

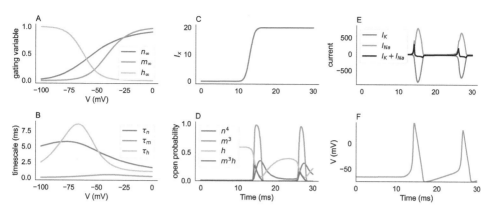

Figure B.1
The HH model. (A, B) Plots of $a_\infty^k(V)$ and $\tau_a(V)$ as a function of V for $a = n, h, m$. (C–F) Simulation in which an input step evokes action potentials. See HodgkinHuxley.ipynb for the code to produce this figure.

and

$$\beta_h = \frac{1}{1 + e^{-0.1(V+35)}}.$$

As before, these can also be written in terms of $m_\infty(V)$, $\tau_m(V)$, $h_\infty(V)$ and $\tau_h(V)$. In general, for all the gating variables, $a = n, h, m$, we have

$$\tau_a(V)\frac{da}{dt} = a_\infty(V) - a$$

where

$$\tau_a(V) = \frac{1}{\alpha_a(V) + \beta_a(V)}$$

and

$$a_\infty(V) = \frac{\alpha_a(V)}{\alpha_a(V) + \beta_a(V)}.$$

The term $a_\infty(V)$ gives the *steady-state* or fixed point of the gating variable $a = n, h, m$ when V is clamped, and $\tau_a(V)$ gives the timescale over which this steady-state value is approached. The dependence of each $a_\infty(V)$ and $\tau_a(V)$ on V is plotted in figure B.1a,b. We can use these plots to think through the membrane currents that we should expect at different values of V.

At rest ($V < -60\,\text{mV}$), the Na channels are essentially closed because $m_\infty(V) \approx 0$ for $V < -60\,\text{mV}$, so $I_{Na} \approx 0$. The K channels are mostly closed at $-60\,\text{mV}$ because $n_\infty(V)$ is just a little bit above zero. But even to the extent that the K channels are open, they only help to keep the membrane potential below $-60\,\text{mV}$ because $E_K = -77\,\text{mV}$, so I_K pulls the membrane potential down. Hence, if V starts below $-60\,\text{mV}$ and we unclamp it, it will stay below $-60\,\text{mV}$ unless extra inward current is applied.

Now consider what would happen if we unclamp V and apply an inward current that is strong enough to bring V toward $-50\,\text{mV}$ or so. The Na channels would start to open very quickly (because $m_\infty(V)$ is no longer so close to zero and $\tau_m(V)$ is very small). This opening of Na channels pulls V up more because $E_{Na} = 50\,\text{mV}$. This creates a positive feedback loop where V increases, which causes $m(V)$ to get larger, which then causes V to increase more, and so on. This positive feedback loop is responsible for the fast upswing of the action potential.

As V increases, h slowly starts to decrease (because $h_\infty(V) \approx 0$ for larger V, but $\tau_h(V)$ is not so small), which slowly shuts down the positive feedback loop (because $h \approx 0$ closes the Na channels). All this time, the K channels have been slowly opening (because $n_\infty(V)$ is larger for larger V and $\tau_n(V)$ is not so small), which works to pull the membrane potential back down (because $E_K = -77\,\text{mV}$), and also slows the positive feedback loop. These combined effects of $n(t)$ and $h(t)$ end the action potential and pull the membrane potential back down toward rest, and even a little below rest.

To see how all this plays out in practice, we need to put all the pieces together into one large model. At the expense of repeating some equations from earlier, the HH model in its entirety is given by

$$C_m \frac{dV}{dt} = -\bar{g}_L(V - E_L) - \bar{g}_K n^4(V - E_K) - \bar{g}_{Na} m^3 h(V - E_{Na}) + I_x(t)$$

$$\frac{dn}{dt} = (1 - n)\alpha_n(V) - n\beta_n(V)$$

$$\frac{dh}{dt} = (1 - h)\alpha_h(V) - h\beta_h(V)$$

$$\frac{dm}{dt} = (1 - m)\alpha_m(V) - m\beta_m(V)$$

(B.2)

where

$$\alpha_n(V) = \frac{0.01(V + 55)}{1 - e^{-0.1(V+55)}}$$

$$\beta_n(V) = 0.125e^{-0.0125(V+65)}$$

$$\alpha_m(V) = \frac{0.1(V + 40)}{1 - e^{-0.1(V+40)}}$$

$$\beta_m(V) = 4e^{-0.0556(V+65)}$$

$$\alpha_h(V) = 0.07e^{-0.05(V+65)}$$

and

$$\beta_h(V) = \frac{1}{1 + e^{-0.1(V+35)}}$$

with V measured in millivolts. Parameter values are $C_m = 1\,\mu\text{F/cm}^2$, $g_L = 0.3\,\text{mS/cm}^2$, $E_L = -54.387$, $\bar{g}_K = 36\,\text{mS/cm}^2$, $E_K = -77\,\text{mV}$, $\bar{g}_{Na} = 120\,\text{mS/cm}^2$, and $E_{Na} = 50\,\text{mV}$. The injected current, $I_x(t)$, models an applied current, such as a current injected by a scientist's electrode.

Note that E_L no longer represents the resting potential of the neuron in the sense that V does not decay toward E_L when $I_x(t) = 0$. Instead, the HH model rests around $V \approx -65\,\text{mV}$ when $I_x(t) = 0$. The leak current, $I_L = -g_L(V - E_L)$, models all the currents not explicitly accounted for in the HH model. Since I_K and I_{Na} affect the resting potential, the membrane potential no longer rests at $V = E_L$. In this sense, the HH model can be considered an extension of the leaky integrator in which the effects of I_K and I_{Na} were pulled *out* of the I_L term and modeled explicitly.

Unlike the leaky integrator model, we cannot write an equation for the solution to equation (B.2). Instead, we can find an approximate, numerical solution to equation (1.6) using the forward Euler method as follows:

```
for i in range(len(time)-1):
  n[i+1]=n[i]+dt*((1-n[i])*alphan(V[i])-n[i]*betan(V[i]))
  m[i+1]=m[i]+dt*((1-m[i])*alpham(V[i])-m[i]*betam(V[i]))
  h[i+1]=h[i]+dt*((1-h[i])*alphah(V[i])-h[i]*betah(V[i]))
  V[i+1]=V[i]+dt*(IL(V[i])+IK(n[i],V[i])+INa(m[i],h[i],V[i])+
    Iapp[i])/Cm
```

where `alphan`, `betan`, and so on are functions defined elsewhere in the code. See `HodgkinHuxley.ipynb` for a full simulation. Figure B.1c–f shows a simulation of the HH model in which action potentials are driven by an increase in the input. The process of action potential generation can be summarized as follows:

1. A positive $I_x(t)$ causes an increase in V.

2. The increase in V causes an increase in m (since m_∞^3 is an increasing function of V), and the increase in m also causes an increase in V (since $E_{Na} = 55$ mV). This creates a positive feedback loop and rapid increase in V. This all happens before h or n can change much because $\tau_m < \tau_h, \tau_n$.

3. After a millisecond or so, h starts to go to zero (because h_∞ is near zero for large V), causing the Na channels to close, which shuts down the positive feedback loop.

4. In the meantime, n has increased (*i.e.*, K channels have opened), so once the Na channels close, the membrane potential is pulled back down to negative values (since $E_K = -77$ mV) as K ions rush out of the cell. This ends the action potential.

5. After h recovers, the cycle repeats.

The mathematical biologist Jim Keener developed an interpretive dance for this process, which he called the "Hodgkin-Huxley Macarena." Search for videos of the dance online. The role of each gating variable in an action potential can be summarized as follows:

- m causes V to increase.

- h stops V from increasing.

- n causes V to decrease.

Importantly, we used a smooth increase to I_x in figure B.1 instead of a sharp step because the HH model can respond in unexpected ways to rapid jumps in $I_x(t)$.

Figure B.2 shows the membrane potential response to multiple pulses of different strengths and to sustained input. Note that the first two pulses were not strong enough to drive an action potential.

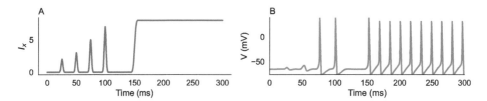

Figure B.2
The HH model driven by pulsatile and sustained inputs. (A) Applied input current and (B) membrane potential of the HH model. See `HHspikes.ipynb` for the code to produce this figure.

Some important concepts related to action potentials are listed here:

• **Threshold:** For an action potential to occur, V needs to get large enough to start the positive feedback loop. The cutoff V to initiate an action potential is called the *threshold potential*, often denoted as V_{th} or θ. When action potentials are driven by an applied current, $I_x(t)$, then the *threshold current* is the value of $I_x(t)$ needed to evoke an action potential. But note that the threshold can depend on the shape of the current used to evoke the action potential. For example, a sharply rising current might evoke an action potential more easily than a slowly rising current. In figure B.2, we can see that there is a threshold pulse strength somewhere between the strengths of the second and third pulses. Can you find a threshold step height for the second half of the simulation?

• **Refractory period:** Directly after an action potential, h is near zero so the Na channels are closed and therefore the neuron can't spike again until the h gates open back up. This "refractory period" lasts for about 2 ms after a spike.

• **All-or-none response:** An action potential either occurs or doesn't. And the magnitude and shape of the action potential are independent of the input that evoked it. These statements are not exactly true because, for example, a miniature action potential can be evoked by an input near the threshold current, but it needs to be very close. An input just a little above or below the threshold will evoke a stereotyped spike or no spike at all. This effect can be seen in figure B.2: Two different input pulses evoke action potentials that are virtually identical in shape.

Limitations of the HH model.   The HH model has several problems that can make it impractical for many purposes. One problem is that its parameters were fit to the dynamics of the squid giant axon. While the basic mechanisms of action potential generation are the same in nearly all types of neurons, the specific parameters and dynamics differ across neuron types. For example, the resting potential for the HH model with $I_x(t) = 0$ is around -65 mV, but it is closer to -72 mV in mammalian cortical neurons. HH-style models that more accurately model mammalian cortical neurons can be developed by changing parameter values and adding other ion channels.

HH-style models also have many complicated properties beyond those studied here. These properties can make it difficult to understand the mechanisms underlying different dynamics observed in simulations.

Perhaps most important, the HH model is computationally expensive to simulate. Small time steps, dt, are needed for accurate simulations of the HH model due to the fast dynamics underlying action potential generation. The use of a smaller time step causes simulations of the HH model to take longer. A larger time step could be used if we used a more advanced ODE solver, but this helps only so much in practice. This computational inefficiency is not an issue when simulating a single neuron over a short time interval, but it can be an issue when simulating networks of many neurons over longer periods of time.

Many of these limitations are resolved, at least partially, by using the exponential integrate-and-fire (EIF) model from section 1.2 or other simplified neuron models such as those described in the next section.

B.2 Other Simplified Models of Single Neurons

B.2.1 Other IF Models

The EIF model introduced in section 1.2 in chapter 1 used a threshold-reset condition to model the recovery of the membrane potential to rest after an action potential. Models of this form are called *integrate-and-fire* models. A more general formulation of IF models is given by

$$\frac{dV}{dt} = f(V, I_x)$$

(B.3)

$$V(t) > V_{th} \Rightarrow \text{spike at time } t \text{ and } V(t) \leftarrow V_{re}.$$

The EIF model is recovered by taking

$$f_{EIF}(V, I_x) = \frac{-(V - E_L) + De^{(V - V_T)/D} + I_x}{\tau_m}.$$

In this subsection, we describe several alternative IF models.

The LIF model. The EIF model is named for the exponential term, $De^{(V-V_T)/D}$, that captures the upswing of an action potential. This upswing of the action potential initiation does not have a large impact on spike timing: If D is small, then the neuron is very likely to spike shortly after V crosses V_T and is very unlikely to spike if V does not cross V_T. As such, we do not lose much by ignoring the action potential initiation and just recording a spike whenever V crosses V_T. This line of reasoning gives rise to the *leaky integrate-and-fire (LIF) model*, which is identical to the EIF model except for being the exponential term:

$$\tau_m \frac{dV}{dt} = -(V - E_L) + I_x(t)$$

(B.4)

$$V(t) > V_T \Rightarrow \text{spike at time } t \text{ and } V(t) \leftarrow V_{re}.$$

In other words, we drop the exponential term from f_{EIF} to get

$$f_{LIF}(V, I_x) = \frac{-(V - E_L) + I_x}{\tau_m}.$$

The LIF model was one of the first neuron models ever developed, being first proposed in 1907 by Marcelle and Louis Lapicque (Lapique 1907; Abbott 1999; Brunel and Van Rossum 2007). As a result, it is sometimes referred to as the *Lapicque model*.

Note that we used V_T in place of V_{th} for the threshold. This is because the soft threshold from the EIF serves as a hard threshold for the LIF, in the sense that a spike occurs

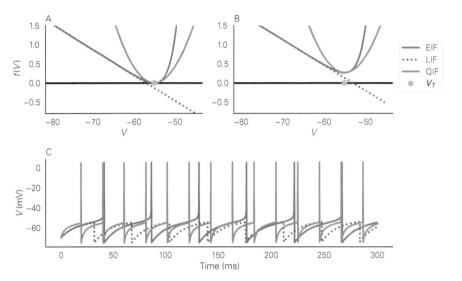

Figure B.3
Comparing the EIF, LIF, and QIF models. (A) The functions $f(V)$ for all three models when $I_x(t) = I_0 = I_{th}$. Parameters for the QIF were chosen to match the second-order Taylor expansion of the EIF at $V = V_T$ (see the text for more details). (B) Same as A, but for a larger value of I_0. (C) Membrane potentials for all three models using the parameters from B. See `EIFLIFQIF.ipynb` for the code to produce this plot.

as soon as the membrane potential enters the region near the onset of an action potential, $V_T \approx -55$ mV. The functions $f_{EIF}(V, I_0)$ and $f_{LIF}(V, I_0)$ are compared in figure B.3a,b and membrane potential traces are compared in figure B.3c. In the limit as $D \to 0$, the spike times of the EIF model converge to those of the LIF model, so the models become effectively equivalent. The LIF model lacks the upward deflection of the membrane potential before the spike times, so the membrane potential traces look less realistic but the spike times are similar.

It is a common practice to perform a *change of coordinates* to simplify the equations of a neuron model. For example, the LIF model is often written as

$$\frac{d\tilde{V}}{d\tilde{t}} = -\tilde{V} + \tilde{I}_x(\tilde{t}) \tag{B.5}$$

$$\tilde{V}(\tilde{t}) > 1 \Rightarrow \text{spike at time } \tilde{t} \text{ and } \tilde{V}(\tilde{t}) \leftarrow \tilde{V}_{re}.$$

This approach is often described as simply taking $E_L = 0$, $\tau_m = 1$, and $V_T = 1$, but a better explanation is that we are changing coordinates from V and t to $\tilde{V} = (V - E_L)/(V_T - E_L)$ and $\tilde{t} = t/\tau_m$. Since this change of coordinates is linear (or, more strictly speaking, affine), we can also view equation (B.5) as a *rescaled* version of equation (1.6) in chapter 1. This also means that we can view the change of coordinates as simply changing the units in which V and t are measured. Specifically, $\tilde{t}$ is measured in units of τ_m (so $t = 3$ should be interpreted as $t = 30$ ms if $\tau_m = 10$), and $\tilde{V}$ measures the distance of V from E_L in a unit of $V_T - E_L$. The advantage of the formulation in equation (B.5) is that it is simpler. The disadvantage, however, is that V and t are no longer represented explicitly in familiar units like mV and ms. But both formulations are equivalent, as exercise B.2.3 shows.

The QIF model. The LIF replaces f_{EIF} with a linear function, which does not model the upswing of the action potential. The *quadratic integrate-and-fire (QIF) model* replaces f_{EIF} with a quadratic function that can still capture the action potential upswing:

$$\frac{dV}{dt} = c(V - V_1)(V - V_2) + I_x(t)$$

$$V(t) > V_{th} \Rightarrow \text{spike at time } t \text{ and } V(t) \leftarrow V_{re}$$

where $c > 0$. In other words, $f_{QIF}(V) = c(V - V_1)(V - V_2) + I_x(t)$. If we want to approximate the EIF with the QIF, we need to relate their parameters in such a way that $f_{QIF}(V, I_x) \approx f_{EIF}(V, I_x)$. It is impossible for the approximation to be accurate at all values of V and I_x. Instead, we can focus on approximating the EIF with time-constant input, $I_x(t) = I_0$, near the bifurcation point given by $I_0 = I_{th}$ and $V = V_T$. This is the point at which the EIF transitions from never spiking to spiking through a saddle-node bifurcation (see section 1.2 and exercise 1.2.1 in chapter 1). The second-degree Taylor polynomial expansion of $F_{EIF}(V, I_0)$ at $V = V_T$ gives the approximation

$$f_{EIF}(V, I_0) \approx \frac{1}{2D\tau_m}(V - V_T)^2 + \frac{I_0 + V_T - E_L + D}{\tau_m}. \tag{B.6}$$

Therefore, if we take $V_1 = V_2 = V_T$, $c = 1/(2D\tau_m)$ and $I_0^{QIF} = (I_0^{EIF} + V_T - E_L + D)/\tau_m$ then the two models behave similarly near the bifurcation point. Indeed, the approximation is accurate when V is near V_T even when I_0 is not near I_{th} for the EIF since equation (B.6) is valid for any value of I_0. Figure B.3 compares the EIF and QIF models with this parameter substitution, and panel (a) shows the case $I_0 = I_{th}$. Note that the EIF and QIF are similar near $V = V_T$, but very different away from $V = V_T$ and the spike times are very different. The values of V_1, V_2, and c can be chosen to better approximate the EIF at different values of V and for different values of I_0, but there is no choice of parameters that provides good approximation at all values of V and I_0.

A linear change of coordinates (or rescaling) of V and t can transform the QIF model to

$$\frac{d\tilde{V}}{d\tilde{t}} = \tilde{V}^2 + \tilde{I}_x. \tag{B.7}$$

When $\tilde{I}_x(\tilde{t}) = \tilde{I}_0$ is constant, the bifurcation between spiking and not spiking occurs at $\tilde{I}_0 = \tilde{I}_{th} = 0$. Hence, $\tilde{I}_x$ and $\tilde{I}_0$ should be interpreted as the excess input above threshold. Since the parameters for the QIF can be chosen to approximate the EIF near the bifurcation point, the EIF can also be rescaled in such a way that equation (B.7) approximates the rescaled EIF near the bifurcation point. Indeed, equation (B.7) is called a "normal form" for a saddle-node bifurcation, meaning that almost any ODE undergoing a saddle-node bifurcation can be rescaled in such a way that it is approximated by equation (B.7) near the bifurcation point. The "almost" qualifier is needed because the statement does not apply to ODEs for which the Taylor series around the bifurcation point has a vanishing quadratic term, but this is unlikely to occur without a symmetry that produces it explicitly.

In this sense, the QIF model approximates a large class of neuron models that transition between spiking and not spiking through a saddle-node bifurcation. However, the usefulness of this approximation is limited by the fact that it applies only near the bifurcation point (near $V = V_T$ and $I_x(t) = I_{th}$). In neuron models, we are often interested in the dynamics away from this bifurcation point. The EIF model does a better job of capturing the phase line of the HH model away from the bifurcation point, so it is often viewed as more accurate (Fourcaud–Trocme et al. 2003; Jolivet, Lewis, and Gerstner 2004; Gerstner et al. 2014). However, the QIF model is more amenable to rigorous mathematical analysis, as we will explore more in section B.2.2.

Refractory periods in IF models. One phenomenon that is not captured by the EIF model or equation (B.3) is the refractory period: After a neuron spikes, it is very unlikely to spike again for the next few milliseconds. Refractory periods arise naturally in the HH model from appendix B.1, but they need to be added explicitly to IF models. An IF with a refractory period can be modeled by adding a refractory condition to equation (B.3) to obtain

$$\frac{dV}{dt} = f(V, I_x)$$

$$V(t) > V_{th} \Rightarrow \text{spike at time } t \text{ and } V(s) = V_{re} \text{ for } s \in [t, t + \tau_{ref}].$$

Here, τ_{ref} is the refractory period, which should be around $1 - 4$ ms. This rule says that the membrane potential should be held at V_{re} for a duration τ_{ref} after a spike. Cortical neurons typically spike only at $1 - 20$ Hz, meaning that the time between two consecutive spikes is typically $50 - 1,000$ ms long. Therefore, a $1 - 4$-ms refractory period will not have a substantial effect on spike timing. But refractory periods might be important when modeling neurons that spike at higher rates or when modeling phenomena where precise spike timing is important.

Adaptive IF models. IF models can also be modified by adding currents that model the effects of active ion channels and other phenomena. A common example is *spike frequency adaptation* (often just called "adaptation" when there is no ambiguity), which is the tendency for consecutive interspike intervals (*i.e.*, the time between consecutive spikes) to get longer over time during sustained spiking (figure B.4a). In other words, the frequency of spiking decreases as the neuron continues to spike. Adaptation can be caused by different mechanisms, but one common mechanism is the presence of hyperpolarizing ion channels that are opened by action potentials and close slowly afterward. The more spikes that occur, the more these ion channels open, making the neuron less excitable after each spike. Adaptation

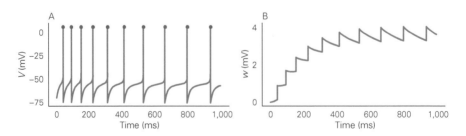

Figure B.4
Simulation of an AdEx model. (A) The membrane potential and (B) the adaptation variable for an AdEx model with time-constant input, $I_x(t) = I_0$. See AdEx.ipynb for the code to produce this plot.

can be modeled by adding an adaptation variable, w, to the EIF model to obtain the *adaptive EIF (AdEx or AdEIF) model*:

$$\tau_m \frac{dV}{dt} = -(V - E_L) + \Delta_T e^{(V - V_T)/\Delta_T} + I_{app}(t) - w$$

$$\tau_w \frac{dw}{dt} = -w + a(V - E_L)$$

$$V(t) > V_{th} \Rightarrow \text{spike at time } t, \; V(t) \leftarrow V_{re}, \text{ and } w \leftarrow w + b$$

where $\tau_w > 0$ determines the timescale of adaptation (around $\tau_w \approx 100 - 1{,}000$ ms), and $a, b \geq 0$ determine the intensity of adaptation.

When the neuron spikes, w is increased by the explicit $w \leftarrow w + b$ rule, by the $a(V - E_L)$ term, or both. Since these two effects are often similar, it is often simpler to set $a = 0$ so w is changed only by the reset rule. An increase in w helps to hyperpolarize V, making it less excitable. When provided with a superthreshold, time-constant input, $I_{app}(t) = I_0$, this can lead to adaptation. Figure B.4 shows a simulation of the AdEx model. The AdEx model is simple and is highly accurate when its parameters are fit to recordings of real neurons, and it has won at least one modeling competition (Brette and Gerstner 2005; Jolivet et al. 2008). For these reasons, some people consider it an ideal model that balances simplicity with biological realism.

Adaptive LIF and QIF models can similarly be defined by adding an adaptation variable. The adaptive QIF model is sometimes called the *Izhikevich model*, named after Eugene Izhikevich, who popularized it. Izhikevich and his colleagues have shown that adaptive QIF models can exhibit a rich diversity of spike-timing dynamics by changing various parameters. For example, the adaptation variable can be made depolarizing by setting $a \leq 0$, $b \leq 0$, or both. This makes the model useful for modeling many different types of neurons with various dynamical properties (Izhikevich 2003, 2004, 2007).

The PIF model.   The EIF, QIF, and adaptive models are slightly more realistic versions of the classic but simple LIF model. However, we can also go in the opposite direction from the LIF: We can make it even simpler. This is achieved by the *perfect integrate-and-fire (PIF)* model, which is the LIF model with the leak conductance removed to get

$$\tau_m \frac{dV}{dt} = I_x(t)$$

$$V(t) > V_{th} \Rightarrow \text{spike at time } t, \; V(t) \leftarrow V_{re}.$$

The membrane potential just integrates the input and spikes when the integral reaches V_{th}. The PIF is an accurate approximation of the LIF when the input current is much larger than the leak, $|I_x(t)| \gg |g_L(V - E_L)|$. The PIF model is not widely employed, but it can be useful for mathematical studies because simple, closed-form equations can be derived for firing rates and other spike train statistics when the input is noisy (Rosenbaum and Josić 2011).

B.2.2 Neural Oscillators and Phase Models
Neural oscillator models.   Recall that IF models with time-constant, superthreshold input $(I_x(t) = I_0 > I_{th})$ have periodic membrane potentials that oscillate around the state space

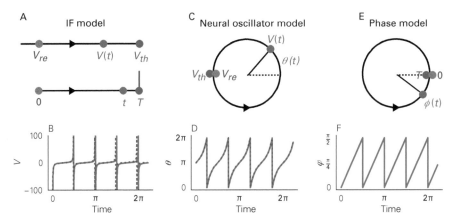

Figure B.5
IF model transformed to an oscillator and phase model. (A) The membrane potential for an IF model with time-constant input is periodic on the interval $[V_{re}, V_{th}]$ and the spike times are periodic on $[0, T_p]$ where T_p is the period. (B) The membrane potential of the QIF from equation (B.9). The solid blue curve is the closed-form solution from equation (B.10). The dashed red curve was obtained using the forward Euler method.(C) An oscillator model is obtained by mapping the state space of the membrane potential onto a circle and measuring the angle, $\theta(t)$, of the state. (D) A simulation of the theta oscillator model obtained from the change of coordinates in equation (B.13) applied to the blue curve in B (solid blue) and by applying the forward Euler method to equation (B.12) (red dashed). (E) A phase model is obtained by keeping track of the amount of time since the last spike. (F) A phase model for the QIF in (A). See the first code cell of `QIFOscPhase.ipynb` for the code to produce these plots.

$[V_{re}, V_{th}]$. The idea behind oscillator models is to wrap the interval $[V_{re}, V_{th}]$ around a circle in such a way that V_{re} and V_{th} become the same point. In this case, the membrane potential, $V(t)$, can be visualized as a point on the circle. The key insight to oscillator models is that it is arguably more natural to keep track of the angle, $\theta(t)$, of the vector that points from the origin to $V(t)$ on the circle. This idea is illustrated in figure B.5a,c. This mapping can work even with time-dependent input, $I_x(t)$, for which spiking is not periodic. The dynamics can be written in terms of the angle

$$\frac{d\theta}{dt} = F(\theta, I_x) \tag{B.8}$$

where $\theta(t) \in [0, 2\pi)$ is the angle in radians. Note that $\theta(t)$ lives on a periodic state space, meaning that any arithmetic on θ needs to be performed modulo 2π. For example, the forward Euler update for equation (B.8) should be

```
theta[i+1]=theta[i]+F(theta[i],Ix[i])*dt
theta[i+1]=np.mod(theta[i+1], 2*np.pi)
```

where the second line takes the modulus with 2π. This ensures that whenever $\theta(t)$ exceeds 2π, it is reset to 0, and when it falls below 0, it is reset to 2π.

If $\theta(t)$ and $F(\theta, I_x)$ are properly derived from a change of coordinates from an IF model, then the oscillator model will be equivalent to the IF model, but with one caveat: In most IF models, the membrane potential can potentially take on the values $V < V_{re}$. In an oscillator model, this would cause the membrane potential to reenter at the threshold, which is not equivalent to the IF model and not biologically realistic. Hence, oscillator models should be used to replace IF models only in situations where the membrane potential rarely or never falls below V_{re}. This assumption is valid if the input is a small perturbation away from superthreshold input ($I_x(t) = I_0 + \epsilon I(t)$ with $I_0 > I_{th}$ and ϵ small), which implies that spiking is nearly periodic.

The QIF is especially amenable to representing as an oscillator model if we take $V_{re} = -\infty$ and $V_{th} = \infty$. This may seem like a strange choice of parameters, but it makes the mathematics work out nicely. To understand this choice, first note that the membrane potential of the QIF model with superthreshold input would reach $V(t) = \infty$ in finite time if we did not have the threshold-reset condition. In other words, it would have a vertical asymptote. Hence, mathematically, it is valid to take $V_{th} = \infty$ because the membrane potential will still reach threshold. In simulations, of course, we cannot actually take $V_{th} = \infty$, but taking a large value of V_{th} achieves similar numerical results. Similarly, if the membrane potential is reset to $V_{re} = -\infty$, it can recover in finite time (another vertical asymptote), but in simulations, we just choose V_{re} to be large and negative. To better see how this works, consider the (rescaled) QIF model:

$$\frac{dV}{dt} = V^2 + I_x(t). \tag{B.9}$$

When $I_x(t) = I_0$, a closed-form solution is given by

$$V(t) = -\sqrt{I_0} \cot\left(t\sqrt{I_0}\right) \tag{B.10}$$

where $V(0) = V_{re} = -\infty$ is the initial condition. More generally, $V(t) = \sqrt{I_0} + \tan(t\sqrt{I_0} + c)$ is the general solution where c is determined by $\tan(c) = V(0)/\sqrt{I_0}$, but for simplicity, we will focus on the solution in equation (B.10). We don't need to write the usual threshold-reset conditions in equation (B.9) because an action potential occurs naturally when the cot term in equation (B.10) has a vertical asymptote: $V(t)$ blows up to ∞ and is then reset to $-\infty$ at the spike times $s_n = n\pi/\sqrt{I_0}$ for $n = 1, 2, \ldots$, so the period of spiking is

$$T_p = \frac{\pi}{\sqrt{I_0}}$$

and the firing rate is $r = 1/T_p = \sqrt{I_0}/\pi$. Figure B.5b compares the closed-form solution from equation (B.10) (solid blue line) to a numerical solution obtained using the forward Euler method with $V_{th}, V_{re} = \pm 100$ (dashed red line). Note that when using the forward Euler method to solve the QIF with large values of V_{re} or V_{th}, you must choose a smaller time step, dt, to prevent large numerical errors caused by large values of $V(t)$.

The QIF with $V_{re} = -\infty$ and $V_{th} = \infty$ is ideal for representing as an oscillator model, in part because of the existence of simple closed-form solutions like equation (B.10), but also because we can never have $V < V_{re}$ when $V_{re} = -\infty$, so we don't need to worry about the caveat mentioned in this discussion. Note that the membrane potential of the EIF would also have a vertical asymptote without the threshold-reset condition, but the EIF does not recover from $V_{re} = -\infty$ in finite time and does not admit simple closed-form solutions, so it is a less amenable oscillator representation than the QIF.

To represent the QIF with $V_{th} = \infty$ and $V_{re} = -\infty$ as an oscillator, we can make a change in coordinates from $V(t) \in [-\infty, \infty]$ to $\theta(t) \in [0, 2\pi)$, satisfying

$$V = \tan(\theta/2). \tag{B.11}$$

The transformed system satisfies

$$\frac{d\theta}{dt} = 1 - \cos(\theta) + (1 + \cos(\theta))I_x(t). \tag{B.12}$$

Spikes in the neuron occur when $\theta(t) = \pi$ since this is where $V = \tan(\theta/2)$ has a vertical asymptote. Note that we can also write the change of coordinates as

$$\theta = 2\tan^{-1}(V) \tag{B.13}$$

but this is ambiguous because the tan function is not invertible. To make this relationship work, you need to define $\tan^{-1}$ to return values on $[0, 2\pi)$. Unfortunately, the `np.arctan` function in NumPy returns values on $[-\pi, \pi)$. To correctly convert from V to $\theta \in [0, 2\pi)$, you can use the following code:

```
theta=2*np.arctan2(1,1/V)
```

When $I_x(t) = I_0$ is constant, equation (B.12) has a saddle-node bifurcation at $I_0 = 0$. Equation (B.12) is called the *theta model* or *Ermentrout-Kopell canonical model*, named after two mathematical neuroscientists, Bard Ermentrout and Nancy Kopell, who derived it as a canonical model of a saddle-node bifurcation on a periodic domain like the circle (Ermentrout and Kopell 1986). This type of bifurcation is sometimes called a *saddle node on an invariant circle (SNIC)*.

Figure B.5d shows a solution to equation (B.12) found by changing coordinates from the closed-form solution in figure B.5b (solid blue line) and from applying Euler's method to equation (B.12) (dashed red line). Note that the solution is continuous despite the jumps in the plot because $0 = 2\pi$ in the periodic domain.

In some ways, oscillator models are a more natural way to represent IF models since the threshold-reset condition is represented naturally. Similarly, a SNIC bifurcation is a more natural way to think about the bifurcation underlying spiking in most IF models. However, as noted previously, oscillators cannot accurately represent IF models when $V(t) < V_{re}$, which limits their applicability.

Phase models. Like oscillator models, *phase models* are motivated by the observation that $V(t)$ is a periodic function when $I_x(t) = I_0 > I_{th}$. But instead of mapping $V(t)$ onto a circle and measuring the angle, we just keep track of the time since the last spike, called the *phase* of the oscillation. Specifically, we can change coordinates to $\phi(t) \in [0, T_p)$, where T_p is the period of the oscillation and $\phi(t)$ is the time that has elapsed since the most recent spike before t. Like $\theta(t)$, $\phi(t)$ also lives on a periodic domain (figure B.5e). While this representation is not very useful when $I_x(t) = I_0$, it can become more so when we include a weak perturbation away from time-constant input, $I_x(t) = I_0 + \epsilon I(t)$.

As an illustrative example, consider the QIF model from equation (B.9) with solution given by equation (B.10). A phase model is obtained by a change of coordinates satisfying

$$V(t) = -\sqrt{I_0}\cot\left(\phi(t)\sqrt{I_0}\right) \tag{B.14}$$

where $\phi(t)$ is restricted to the periodic domain, $\phi(t) \in [0, T_p)$, determined by the period $T_p = \pi/\sqrt{I_0}$ of the QIF. Notice that equation (B.14) is just obtained by plugging $\phi(t)$ in for t in the solution from equation (B.10). This is a general approach for changing coordinates to a phase model because it ensures that $\phi(t) = t$ when $t \in [0, T_p)$. More generally, we have

$$\frac{d\phi}{dt} = 1.$$

Since $\phi \in [0, T_p)$ lives on a periodic domain, this means that

$$\phi(t) = \mathrm{mod}(t, T_p) \tag{B.15}$$

with spikes occurring whenever $\phi(t) = 0$. This model is not very useful by itself because it can't tell us anything that we don't already know about the QIF.

To show how phase models can be useful, let us generalize to a QIF with time-dependent input. Specifically, consider the QIF with $V_{th} = \infty$, $V(0) = V_{re} = -\infty$, and a delta function pulse delivered at some time $t = \phi_0 \in [0, T_p)$:

$$\frac{d\hat{V}}{dt} = \hat{V}^2 + I_0 + a\delta(t - \phi_0). \tag{B.16}$$

We use the notation $\hat{V}(t)$ to distinguish the perturbed solution from the unperturbed solution.

Figure B.6a shows the membrane potential of a perturbed ($\hat{V}(t)$; red line) and unperturbed ($V(t)$; blue line) QIF. Notice the small jump at time $t = \phi_0$, which causes the spike to happen earlier in the perturbed QIF. After time $t = \phi_0$, the perturbed QIF follows the same trajectory as the unperturbed QIF, except that it is advanced in time. In other words,

$$\hat{V}(t) = V(t + \Delta\phi)$$

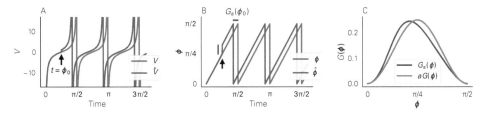

Figure B.6
Phase resetting for the QIF model. (A) The unperturbed (blue) and perturbed (red) membrane potentials satisfying equations (B.9) and (B.16), respectively, with $V(0) = V_{re} = -\infty$. The perturbation is a pulse of amplitude a delivered at time $t = \phi_0$. (B) The phase representation of the membrane potentials. The pulse causes the phase to jump by an amount given by the PRC, $G_a(\phi_0)$, which advances the next spike by the same amount (purple lines, which are the same length). (C) The PRC ($G_a(\phi)$; purple) and the infinitesimal PRC ($aG(\phi)$; green) as a function of ϕ. See the second code cell of QIFOscPhase.ipynb for code to produce these plots.

for $t > \phi_0$ where $\Delta\phi$ is called a *phase shift*. Note that all spike times in the perturbed QIF are advanced by $\Delta\phi$ (occurring sooner if $\Delta\phi > 0$, and later if $\Delta\phi < 0$). The phase shift depends on the amplitude, a, of the perturbation, and on the phase, ϕ_0, at which it was delivered, so we define the function

$$G_a(\phi_0) = \Delta\phi$$

to measure the phase shift for given values of a and $\phi_0 \in [0, T_p)$. The function $G_a(\phi)$ is called the *phase-resetting curve (PRC)* for the QIF model. It is perhaps more natural to talk about PRCs using a phase model. The unperturbed phase, $\phi(t)$, satisfies equations (B.14)–(B.15). The perturbed phase jumps up by $\Delta\phi = G_a(\phi_0)$ at the perturbation, which is equivalent to shifting the phase in time by the same amount; that is, the purple lines in figure B.6b has the same length. Specifically, the perturbed phase satisfies

$$\frac{d\hat{\phi}}{dt} = 1 + G_a(\phi_0)\delta(t - \phi_0).$$

The PRC can be calculated in closed-form by noting that $\Delta\phi$ measures the amount of time for $V(t)$ to evolve from $V(\phi_0)$ to $V(\phi_0 + a)$. In other words, right after the pulse, we have $\hat{V}(\phi_0) = V(\phi_0) + a$, and we also have $\hat{V}(\phi_0) = V(\phi_0 + \Delta\phi)$, which implies that

$$V(\phi_0 + \Delta\phi) = V(\phi_0) + a. \tag{B.17}$$

Using the closed-form solution from equation (B.10), we can solve for $\Delta\phi = G_a(\phi_0)$ to get

$$G_a(\phi) = \frac{1}{\sqrt{I_0}} \tan^{-1}\left(\frac{\sqrt{I_0}\tan\left(\sqrt{I_0}\phi\right)}{\sqrt{I_0} - a\tan\left(\sqrt{I_0}\phi\right)}\right) \tag{B.18}$$

which is plotted in figure B.6c (purple line).

The QIF is somewhat special in that it has a relatively simple closed-form equation for $V(t)$, which we can use to derive a closed-form equation for the PRC. In other situations, the PRC cannot easily be derived in closed-form, but we can instead use the *infinitesimal PRC*:

$$G(\phi) = \lim_{a \to 0} \frac{G_a(\phi)}{a} = \left.\frac{\partial G_a(\phi)}{\partial a}\right|_{a=0}. \tag{B.19}$$

The infinitesimal PRC can be used to approximate phase differences when a is small, using

$$\Delta\phi = G_a(\phi) \approx aG(\phi).$$

For the QIF model, we can derive $G(\phi)$ directly from equation (B.18) to get

$$G(\phi) = \frac{1}{\sqrt{I_0}} \sin^2(\sqrt{I_0}) \tag{B.20}$$

which is plotted in figure B.6c (green line). For other models, the infinitesimal PRC can often be derived or measured more easily than the true PRC.

Phase models are useful for studying networks of weakly coupled neurons. As a simple example, consider a network in which one QIF neuron spikes periodically and provides pulsatile synaptic input to another neuron:

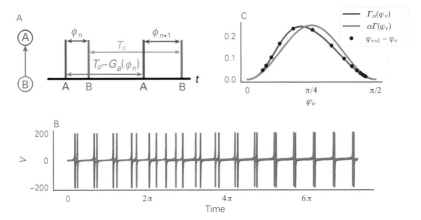

Figure B.7
Using a coupled phase oscillator model to analyze a QIF receiving pulsatile input from another QIF. (A)
QIF neuron A (red) receives pulsatile input from QIF neuron B (blue), as defined by equation (B.21). The phase
difference, ψ_n, measures the time from the nth spike in neuron A (red bars) until the next spike in neuron B (blue
bars). Neuron B spikes periodically with period T_p, while the next spike in neuron A is advanced by an amount
$G_w(\psi_n)$, so the time between consecutive spikes in A is $T_p - G_w(\psi_n)$. (B) A simulation of the model shows spikes
in A advancing (occurring sooner) until they become nearly synchronous with the previous spike in B. (C) The
relationship between $\psi_{n+1} - \psi_n$ and ψ_n (black dots from simulation) is described by the PRC (purple line) and
approximated by the infinitesimal PRC (green line), according to equation (B.22). See the second code cell of
`QIFOscPhase.ipynb` for the code to produce these plots.

$$\frac{dV_A}{dt} = V_A^2 + I_0$$

$$\frac{dV_B}{dt} = V_B^2 + I_0 + w\delta(V_A - V_{th})$$

(B.21)

with $V_{th} = \infty$ and $V_{re} = -\infty$, but the initial conditions are arbitrary and not necessarily
equal ($V_A(0) \neq V_B(0)$ in general), so the neurons do not necessarily spike at the same time.
The delta function term should be interpreted as a kick delivered to V_B each time that V_A
spikes. Changing to phase coordinates, $V_X = -\sqrt{I_0}\cot\left(\phi_X\sqrt{I_0}\right)$, we can write

$$\frac{d\phi_A}{dt} = 1$$

$$\frac{d\phi_B}{dt} = 1 + w\sum_n G(s_n^A)$$

where s_n^A is the nth spike time of neuron A; that is, $\phi_A(s_n^A) = 0$. This system is called a
periodically forced oscillator because V_B is an oscillator and the perturbation from spikes
in neuron A are periodic. The model was first studied in 1964 in the context of pacemaker
neurons that spike almost periodically (Perkel et al. 1964).

Now let's measure the time, ϕ_n, from the nth spike in neuron B to the next spike in neuron
A. This is called the *phase difference* between the neurons. From the diagram in figure B.7a,
you can see that

$$T_p + \psi_n = T_p - G_w(\psi_n) + \psi_{n+1}$$

which can be solved to obtain

$$\psi_{n+1} = \psi_n + G_w(\psi_n). \tag{B.22}$$

This equation tells us how the phase difference evolves over time. In deriving this equation, we have implicitly assumed that there is exactly one spike in neuron B between every pair of spikes in neuron A (*i.e.*, the neurons spike consecutively), but this approach can be generalized to situations where this assumption is not true. Equation (B.22) is called a *map* of the state space $[0, T_p)$. It can easily be used to compute all ψ_n starting from any initial ψ_0. A map can be viewed as a discrete-time differential equation if we interpret n as discrete time. Fixed points of this map satisfy

$$\psi^* = \psi^* + G_w(\psi^*)$$

which is equivalent to

$$G_w(\psi^*) = 0.$$

For the QIF PRC, there is a unique fixed point at $\psi^* = 0$ or, equivalently, $\psi^* = 2\pi$. Moreover, the PRC is strictly positive for $\phi \neq 0$, so the phase increases monotonically toward the stable fixed point at $\psi^* = 2\pi$. In other words, the phase difference between a spike in A and the next spike in B approaches 2π, so the spike in A becomes nearly synchronous with the *previous* spike in B. In other words, neuron B spikes right before neuron A in the steady state.

This approach can be generalized to networks of recurrently coupled oscillators. The literature on oscillators and phase models is vast and rich, with numerous applications beyond neuroscience (Perkel et al. 1964; Kopell and Ermentrout 1990; Strogatz and Stewart 1993; Hoppensteadt and Izhikevich 1997; Oprisan, Prinz, and Canavier 2004; Ermentrout and Terman 2010; Canavier and Achuthan 2010; Stiefel and Ermentrout 2016). In neuroscience, oscillator models are useful for modeling neurons that spike nearly periodically, which is not uncommon outside the cerebral cortex or even outside the brain altogether, but they are arguably less useful for modeling cortical neurons, which spike irregularly and often satisfy $V(t) < V_{re}$.

Exercise B.2.1. Consider the LIF with time-constant input, $I_x(t) = I_0$. Compute the threshold input, I_{th} for which the neuron spikes if $I_0 > I_{th}$ and does not spike if $I_0 < I_{th}$. Compare to the threshold that you computed for the EIF model in Exercise 1.2.1.

Exercise B.2.2. For the EIF and LIF curves from figure B.3, experiment with different values of D to see how it affects the relationship between the EIF and LIF.

Exercise B.2.3. The change of coordinates that gives rise to equation B.34 should satisfy the following:

If $V(t)$ satisfies equation (B.4), $\tilde{V}(\tilde{t})$ satisfies equation (B.5), and $V(0) = (\tilde{V}(0) - E_L)/(V_T - E_L)$, then $V(t) = (\tilde{V}(\tilde{t}) - E_L)/(V_T - E_L)$ for all t where $\tilde{t} = t/\tau_m$.

Derive the values of $\tilde{I}(\tilde{t})$ and $\tilde{V}_{re}$ that make this statement true.

Exercise B.2.4. For the QIF curves in figure B.3, experiment with different values of time-constant input, $I_x(t) = I_0$, to see how that changes the relationship among the three models.

Try to find parameters for the QIF model that give a better approximation to the EIF model.

Exercise B.2.5. How could we implement a refractory period when simulating an EIF or a network of EIF models using Euler's method? Try modifying `EIF.ipynb` to include refractory periods.

Exercise B.2.6. Show that equation (B.10) satisfies equation (B.9).

Exercise B.2.7. Show that $\theta(t)$, as defined in equation (B.11), satisfies equation (B.12) whenever $V(t)$ satisfies equation (B.9). You will need to use some trigonometric identities.

Exercise B.2.8. One advantage of the theta model is that you can use larger values of dt without losing numerical accuracy. Increase dt in figure B.5d, and then compare how the accuracy of the forward Euler method differs between the solutions for $V(t)$ and $\theta(t)$.

Exercise B.2.9. The infinitesimal PRC generally satisfies (Ermentrout and Terman 2010)

$$G(\phi) = \frac{1}{V'(\phi)}. \tag{B.23}$$

Show that equation (B.23) also gives equation (B.20) by plugging in $V'(\phi) = V(\phi)^2 + I_0$, and then plugging in equation (B.10) and simplifying. Equation (B.23) is derived from the adjoint dynamics of $V(t)$ (don't worry if you don't know what "adjoint" means in this context—it is explained by Ermentrout and Terman (2010)), so the infinitesimal PRC is sometimes also called the *adjoint PRC*. Note that equation (B.23) can be used to easily derive the phase shift, $\Delta\phi$, as a function of the *membrane potential*, $V(\phi)$, at the time of the perturbation, even when a closed-form equation for $V(t)$ is not known. If you want a challenge, try deriving equation (B.23) by combining equations (B.17) and (B.19).

Exercise B.2.10. For the simulation in figure B.7, what is the first phase difference, ψ_1? Use this to compute the next several phase differences using equation (B.22) and compare your results to the simulations. Now try reproducing figure B.7b,c with inhibitory coupling ($w < 0$). Do the neurons still synchronize? Does neuron B still spike just before neuron A in the steady state? Note that small errors accumulate, so you should use $V_{th}, V_{re} = \pm 200$ and $dt = 0.0001$ to get accurate simulations.

B.2.3 Binary Neuron Models

Binary neuron models offer a drastic simplification of IF models by ignoring membrane potential dynamics entirely and just mapping inputs to spikes. In binary neuron models, the state of each neuron is binary; that is, it is either spiking or not. We ignore the membrane potential and just model a mapping from inputs directly to the binary spiking state.

The first binary neuron models were developed by Walter Pitts, Warren McCulloch, and Jerome Lettvin in the 1940s and 1950s (McCulloch and Pitts 1943). Their model was similar to a single-layer feedforward artificial neural network (ANN), but with binary inputs and activations. Specifically, the *McCulloch-Pitts model* model can be described by

$$S = H(W_x S^x - \theta)$$

where $H(\cdot)$ is the Heaviside function, θ is a threshold, $\boldsymbol{S}^x$ is an N_x-dimensional vector of presynaptic inputs, W_x is a $1 \times N_x$ feeedforward connectivity matrix (really, just a row vector), and S is an output. This is equivalent to a rate network model of a single neuron with a Heaviside f-I curve and feedforward input $I = W_x \boldsymbol{S}^x$.

The presynaptic inputs and the output are binary, $\boldsymbol{S}_j^x, S \in \{0, 1\}$, and are meant to represent the state of postsynaptic (S) and presynaptic ($\boldsymbol{S}^x$) neurons where $S = 1$ represents a spiking state and $S = 0$ represents a nonspiking state. The weights, W_x, can be designed to perform logical operations on $\boldsymbol{S}^x$. Pitts, McCulloch, and Lettvin hoped that the weights could be trained to model human like cognition, memory, and reasoning. However, the model can implement only simple functions of binary inputs. While the McCulloch-Pitts model is important for historical reasons and it can be viewed as a predecessor to powerful, modern ANNs, it is not widely used today.

Modern binary network models are similar to the McCulloch-Pitts model, but with time-dependent states and recurrent connections. Binary models are typically studied with discrete time, $n = 1, 2, \ldots, T$ instead of continuous time. Specifically, consider a recurrent network of N neurons and let $\boldsymbol{S}(n)$ be the N-dimensional vector of spike trains at time step n. Then

$$\boldsymbol{S}_j(n) \in \{0, 1\}$$

where $\boldsymbol{S}_j(n)$ is the state of neuron j at time step n with 1 corresponding to spiking and 0 corresponding to not-spiking or silent. The vector of inputs at time step n is defined as

$$\boldsymbol{I}(n) = W\boldsymbol{S}(n) + \boldsymbol{X}(n)$$

where W is a recurrent connectivity matrix and $\boldsymbol{X}(n)$ is the external input. There are two common conventions for updating the spike trains. The first is an *asynchronous* update in which a single neuron, j, is chosen randomly and then updated according to

$$\boldsymbol{S}_j(n+1) = H(\boldsymbol{I}_j(n) - \theta) = \begin{cases} 1 & \boldsymbol{I}_j(n) \geq \theta \\ 0 & \boldsymbol{I}_j(n) < \theta \end{cases}$$

or they can be updated all at once, which is called a *synchronous update*:

$$\boldsymbol{S}(n+1) = H(\boldsymbol{I} - \theta).$$

You can also select a random subset of neurons to update on each time step. An alternative formulation of binary networks uses $\boldsymbol{S}_j(n) = -1$ for the silent state, which makes binary neuron models equivalent to "spin glass" models studied in physics. The two formulations can be made equivalent by modifying θ and $\boldsymbol{I}^x$. Binary neuron models capture the opposite limit from the PIF model: They roughly approximate the spiking dynamics of an LIF model when the leak is very strong (g_L large or, equivalently, τ_m small).

While binary neuron models are a rough abstraction from real neurons, they have provided a lot of insight into the dynamics of recurrent networks of neurons. Networks of binary neurons were the first neural network models for which mean-field theories were extensively developed, borrowing techniques from statistical physics (Hopfield 1982; Amit, Gutfreund, and Sompolinsky 1985; Ginzburg and Sompolinsky 1994; Vreeswijk and Sompolinsky

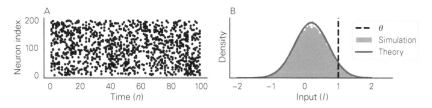

Figure B.8
Raster plot and input distribution for a binary neural network. (A) Raster plot with a dot each time a neuron is in a spiking state. (B) Histogram of inputs (gray; normalized to represent a density) compared to the density estimated by the mean-field theory in equation (B.28) using the empirical firing rate for r. The dashed black line shows the threshold, θ. See `BinaryNet.ipynb` for the code to produce these plots.

1998). As a simple demonstration, consider a binary neural network with

$$W_{jk} \sim \mathcal{N}(\mu_w, \sigma_w)$$

$$X_j(n) \sim \mathcal{N}(\mu_x, \sigma_x)$$

where $z \sim \mathcal{N}(\mu, \sigma)$ should be interpreted to mean that z follows a normal (or Gaussian) distribution with mean μ and standard deviation σ, and where all W_{jk} and $X_j(n)$ are assumed independent. Figure B.8a shows a raster plot from a simulation of this model. Interestingly, the distribution of total inputs, $I_j(n)$, appears to approximately follow a normal distribution (figure B.8b).

To understand why $I_j(n)$ might look approximately normally distributed, consider the total input to a single neuron:

$$I_j(n) = \sum_{k=1}^{N} W_{jk} S_k(n) + X_j(n). \tag{B.24}$$

Note that $X_j(n)$ are normally distributed by assumption. The $S_k(n)$ obey a Bernoulli distribution, but so long as they are approximately independent from each other and from the W_{jk} terms, then the central limit theorem tells us that the sum in equation (B.24) is approximately normally distributed. Moreover, if $S_k(n)$ are approximately independent from $X_j(n)$, then $I_j(n)$ is approximately normally distributed. These independence assumptions are difficult to justify rigorously, and they do not hold in all parameter regimes, but the mean-field theory can proceed by assuming that they are valid.

A naive mean-field theory (analogous to the approach in section 2.3 in chapter 2 for spiking networks) would take expectations of $I_j(t)$ and making the approximation that $r = H(\bar{I} - \theta)$, but this would give a firing rate of 0 or 1, which is inaccurate. An accurate mean-field theory requires us to account for the variability in $I_j(n)$ across time, n. This approach is sometimes called *stochastic mean-field* theory. Let us first define the expectation of $I_j(n)$ over n:

$$\mu_j = E_n[I_j(n)] = \sum_{k=1}^{N} W_{jk} r_k + \mu_x \tag{B.25}$$

where

$$r_k = E_n[S_k(n)]$$

is the firing rate of neuron j. Similarly, we can take the variance across n to get

$$\sigma_j^2 = \text{var}_n(\boldsymbol{I}_j(n)) = \sum_{k=1}^{N} W_{jk}^2 \boldsymbol{r}_k(1 - \boldsymbol{r}_k) + \sigma_x^2 \tag{B.26}$$

where we used the fact that $\boldsymbol{S}_k(n)$ is a Bernoulli random variable to derive its variance. We can then make the approximation that $\boldsymbol{I}_j(n)$ follows a normal distribution for each j. If this is the case, then the firing rate of neuron j is equal to the proportion of time that $\boldsymbol{I}_j(n)$ spends above threshold (see the black dashed line in figure B.8b):

$$r_j = E_n[H(\boldsymbol{I}_j(n)) - \theta] = 1 - F(\theta; \boldsymbol{\mu}_j^I, \boldsymbol{\sigma}_j^I) \tag{B.27}$$

where

$$F(x; \mu, \sigma) = \int_{-\infty}^{x} \frac{1}{\sigma\sqrt{2\pi}} e^{-(x-\mu)^2/(2\sigma^2)}$$

is the cumulative distribution function of the normal distribution, which can be computed in Python as follows:

```
from scipy.stats import norm
y=norm.cdf(x,loc=mu,scale=sigma)
```

Equations (B.25), (B.26), and (B.27) give a *heterogeneous, stochastic mean-field theory* because they describe the statistics of each neuron individually (hence representing the heterogeneity of statistics in the network). This can help quantify the variability across neurons inherited from the quenched variability in W, and it can also be applied when the external input statistics, μ_x and σ_x, vary across neurons. When the network is statistically homogeneous, like the one considered here, a simpler stochastic mean-field theory is obtained by taking expectations over neurons, j, to average out quenched variability and obtain

$$\mu_I = N\mu_w r + \mu_x$$

$$\sigma_I^2 = N\mu_w^2 r(1 - r) + \sigma_x^2 \tag{B.28}$$

$$r = 1 - F(\theta; \mu_I, \sigma_I)$$

where μ_I, σ_I^2, and r approximate the average input mean, input variance, and firing rate in the network. Note that this derivation is not completely rigorous. We had to make several independence assumptions (e.g., assuming that $\boldsymbol{S}_k(n)$, W_{jk}, and $\boldsymbol{X}_k(n)$ are independent) and also had to pass expectations inside nonlinear functions (e.g., assuming that $E[F(\theta; \boldsymbol{\mu}_j^I, \boldsymbol{\sigma}_j^I)] = F(\theta; \mu_I, \sigma_I)$). Some of these approximations can be justified with more detailed calculations, and some of them do not hold in all parameter regimes. Nevertheless, equation (B.28) represents an approximation that we can test empirically. A simple way to test equation (B.28) is to compute r empirically from simulations:

```
r=np.mean(S)
```

and then plug this value of r into equation (B.28) to compute μ_I and σ_I. These values can then be used to approximate the probability density function of $\boldsymbol{I}_j(n) \sim \mathcal{N}(\mu_I, \sigma_I)$. The red curve in figure B.8b shows the result of this approximation, which agrees well with the empirical distribution. Then the consistency of the approximation can be checked further by computing $r_{mf} = 1 - F(\theta; \mu_I, \sigma_I)$ and comparing its value to `r=np.mean(S)`.

The stochastic mean-field theory of binary networks has been developed much more deeply than the relatively simple approximation described here, and it has parallels in statistical physics (Hopfield 1982; Amit, Gutfreund, and Sompolinsky 1985; Ginzburg and Sompolinsky 1994; Vreeswijk and Sompolinsky 1998; Renart et al. 2010).

Exercise B.2.11. In figure B.8, compute $r_{mf} = 1 - F(\theta; \mu_I, \sigma_I)$ and compare it to `r=np.mean(S)`.

Exercise B.2.12. Note that computing μ_I, σ_I, and r_{mf} in Exercise B.2.11 still requires a simulation since we need to first compute `r=np.mean(S)`. An alternative approach is to numerically search for solutions (μ_I, σ_I, and r) that satisfy equation (B.28). Try this: You can write your own numerical solver, or you can use `scipy.optimize.fsolve`.

B.3 Conductance-Based Synapse Models

In section 1.3 of chapter 1, we defined a model of synapses in which each presynaptic spike from a given presynaptic neuron evokes a characteristic synaptic current waveform. In particular, synaptic currents were modeled by

$$I_a(t) = J_a \sum_j \alpha_a(t - s_j^a)$$

where the $\alpha_a(t)$ is postsynaptic current (PSC) waveform. This is called a *current-based synapse model* because each presynaptic spike directly evokes a current. In reality, presynaptic spikes do not evoke currents directly. The neurotransmitter molecules released by a presynaptic spike open ion channels that change the *conductance* of the postsynaptic neuron's membrane to specific types of ions. The synaptic current comes from the resulting flow of ions through these channels. A more detailed model of this process is given by the *conductance-based synapse model*:

$$C_m \frac{dV}{dt} = -g_L(V - E_L) - g_e(t)(V - E_e) - g_i(t)(V - E_i)$$

$$g_e(t) = J_e \sum_n \alpha_e(t - s_n^e) \tag{B.29}$$

$$g_i(t) = J_i \sum_n \alpha_i(t - s_n^i).$$

where $g_a(t)$ is the *synaptic conductance*, $\alpha_a(t)$ is the *postsynaptic conductance (PSC)* waveform, and E_a is the *synaptic reversal potential*. (Confusingly, the abbreviation "PSC" is often used to refer to "postsynaptic current" and "postsynaptic conductance," and the difference often needs to be inferred from the context). Note that for conductance-based synapse models, we can still define the synaptic currents as follows:

$$I_e(t) = -g_e(t)(V - E_e)$$

$$I_i(t) = -g_i(t)(V - E_i)$$

but they are not identical to the currents produced by the current-based model in equation (1.11) in chapter 1.

To understand how conductance-based synapse models work, first consider a single presynaptic excitatory spike under the model in equation (B.29), which causes a transient increase in $g_e(t)$. During this increase, $V(t)$ is pulled toward the reversal potential, E_e. Similarly, an inhibitory presynaptic spike transiently pulls the membrane potential toward E_i.

The polarity (excitatory or inhibitory) of a synapse is determined by the reversal potential, which is determined by the type of neurotransmitter released. The most common type of excitatory neurotransmitter in the cortex is *glutamate*, which has a reversal potential near $E_e = 0$ mV, so excitatory spikes pull the membrane potential toward 0 mV. Since the membrane potential is usually closer to -70 mV, this essentially always results in a positive (*i.e.*, inward or depolarizing) current. The most common type of inhibitory neurotransmitter in the cortex is *gamma-aminobutyric acid (GABA)*. The reversal potential of GABAergic synapses actually changes during development (as an animal grows from the womb to adulthood), but in developed mammals, it is around $E_i = -75$ mV.

The shape and timescale of PSC waveforms, $\alpha_a(t)$, is determined partly by the type of receptor on the postsynaptic neuron's membrane. A common type of receptor for glutamate are *alpha-amino-3-hydroxy-5-methyl-4-isoxazolepropionic acid (AMPA)* receptors. A common GABA receptor type is $GABA_B$. The decay timescales of AMPA and $GABA_B$ receptors are around $\tau_s \approx 3 - 10$ ms, with AMPA a little bit faster than $GABA_B$.

Note that J_e and J_i have different units for conductance-based synapse models than they do for current-based models. Indeed, in conductance-based synapse models, J_i is *positive* since conductance, $g_i(t)$, is always positive. The inhibitory nature of the inhibitory synapses comes from the fact that $V - E_i$ is typically positive, so $I_i(t) = -g_i(t)(V - E_i)$ is typically negative. This is different from current-based models for which negative currents are produced by negative synaptic weights, $J_i < 0$.

If we use the exponential model from equation (1.8) for $\alpha_a(t)$, then equation (B.29) can be rewritten as

$$\tau_m \frac{dV}{dt} = -(V - E_L) - g_e(t)(V - E_e) - g_i(t)(V - E_i) + I_x(t)$$

$$\tau_e \frac{dg_e}{dt} = -g_e + J_e S_e \qquad\qquad (B.30)$$

$$\tau_i \frac{dg_i}{dt} = -g_i + J_i S_i$$

analogous to the current-based model in equation (1.11). Figure B.9 shows a simulation of the model in equation (B.30). Notice that this conductance-based model produces similar membrane potential dynamics to the current-based model in figure 1.5. However, as the following exercises show, the two models can behave differently in some scenarios.

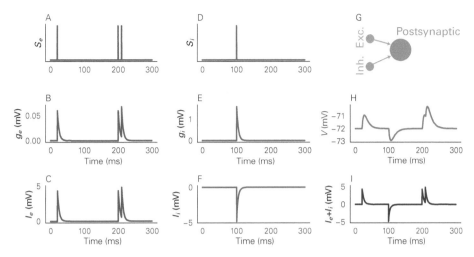

Figure B.9
A leaky integrator model driven by two conductance-based synapses. (A) Excitatory spike density. Each vertical bar represents a spike in the excitatory presynaptic neuron, modeled as a Dirac delta function. (B, C) Excitatory synaptic conductance (B) and current (C) generated by the spikes in (A) using an exponential synapse model. (D–F) Same as (A–C), but for an inhibitory synapse. (G) Schematic of an excitatory and inhibitory neuron connected to a postsynaptic neuron. (H) Membrane potential of the postsynaptic neuron modeled as a leaky integrator. (I) Total synaptic current of the neuron. See `ConBasedSynapses.ipynb` for the code to produce this figure.

Exercise B.3.1. Inhibitory shunting. Since $I_i(t) = -g_i(t)(V - E_i)$, the magnitude of the inhibitory synaptic current depends on the distance of the membrane potential from the synaptic reversal potential. If $V(t) = E_i$ when an inhibitory spike arrives, the spike will not cause a postsynaptic response. If $V(t)$ is very close to E_i, the response will be very small. Simulate a leaky integrator with an inhibitory synapse and a single inhibitory presynaptic spike. Use the initial condition $V(0) = V_0$ and a time-constant external input, $I_x(t) = I_0 = V_0 - E_L$, to implement a "holding potential" at V_0; that is, a potential at which $V(t)$ is held until the spike arrives. Observe how the height of the PSP changes with the value of V_0. What happens when V_0 is close to $E_i = -75$ mV? What happens when you take $V_0 = E_i$ mV? This effect is called "shunting." Shunting can be a strong effect for inhibitory synapses with $E_i = -75$ mV, but it is typically unnoticeable for excitatory synapses with $E_e = 0$ mV. Why?

Exercise B.3.2. Repeat the simulations from Exercise B.3.1 with the current-based synapse model from equation (1.7). Does the model produce shunting? Why or why not?

B.4 Neural Coding

In section 2.1 of chapter 2, we saw how the orientation of a drifting grating stimulus was encoded in the firing rate of a visual cortical neuron. In this section, we develop methods for decoding the orientation of a drifting grating stimulus from recordings of neurons. Effectively, we are developing methods for reading an animal's brain!

Decoding a single neuron. For illustrative purposes, let's begin by decoding orientation from a single neuron before moving to a population of neurons. Figure B.10a shows a

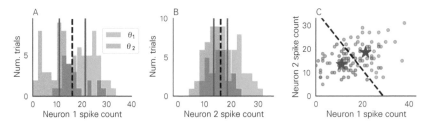

Figure B.10
Decoding spike counts of two visual cortical neurons recorded under two stimuli. (A) A histogram of spike counts recorded from a neuron in a monkey's visual cortex during the presentation of drifting grating stimuli with two different orientations, θ_1 = (red) and θ_2 = (blue). The vertical red and blue lines show the mean spike counts under each stimulus, and the dashed line shows the decision boundary, z. The decision algorithm (B.34) guesses θ_1 if a spike counts is larger than z, and guesses θ_2 if it's smaller. (B) Same as (A), but for a different neuron. (C) The spike counts of both neurons under both conditions. Each dot is one trial. The stars show the mean spike counts under each condition and the dashed line is the decision boundary for the decision algorithm in (B.37). See the first code cell of `PopulationCoding.ipynb` for the code to produce this figure.

histogram of spike counts of a single neuron recorded from a monkey's visual cortex while the monkey watched drifting grating stimuli at two orientations, $\theta_1 = 120°$ and $\theta_2 = 150°$, for fifty trials each (same as figure 2.3b in section 2.1, but a different neuron and different orientations). Looking at this histogram, we can see that if a neuron spiked more than thirty times during a trial, it would be a safe bet that the orientation was θ_1. If the neuron spiked fewer than five times, it would be a safe bet that the orientation was θ_2. But spike counts near fifteen are more ambiguous. Given a spike count, how should we guess the orientation, and how accurate will we be?

Let us abstract away from the data first and consider a statistical model. Let N be a random variable representing the spike count on a given trial and define

$$P(N|\theta_1) = \text{Pr}(\text{spike count} = N \mid \text{orientation} = \theta_1)$$

to be the probability distribution of spike counts under orientation θ_1 and similarly for $P(N|\theta_2)$. These are the distributions estimated by the histograms in figure B.10a. For neural decoding, we care about the opposite distributions: the probability of θ_1 or θ_2 given a spike count. These can be computed using Bayes's theorem:

$$P(\theta_1 \mid N) = \frac{P(N \mid \theta_1)P(\theta_1)}{P(N)}$$
$$P(\theta_2 \mid N) = \frac{P(N \mid \theta_2)P(\theta_2)}{P(N)}. \tag{B.31}$$

These are called the *likelihoods* of θ_1 and θ_2. If we could only see N, then our best guess of θ would be to guess θ_1 when θ_1 is more likely than θ_2—that is, when

$$P(\theta_1 \mid N) > P(\theta_2 \mid N).$$

In many cases (including the data from figure B.10), θ_1 and θ_2 occur with equal probability, $P(\theta_1) = P(\theta_2) = 0.5$. Combining this with equation (B.31), we find the following algorithm:

$$\text{Guess } \theta_1 \text{ whenever } P(N \mid \theta_1) > P(N \mid \theta_2). \tag{B.32}$$

This is an optimal decision algorithm in the sense that it maximizes the probability of guessing correctly under the assumptions made here, but it is useful only if we know $P(N \mid \theta_1)$ and $P(N \mid \theta_2)$.

The only thing left to do is to estimate $P(N \mid \theta_1)$ and $P(N \mid \theta_2)$. It might be tempting to use the raw data to estimate the probabilities. For example, since $N = 3$ occurred in five of the fifty trials during which θ_1 was presented, we could take $P(N = 3 \mid \theta_1) = 0.1$. But this approach would overfit the noise in the trials. Instead, we should use a *parameterized distribution* to represent $P(N \mid \theta_j)$. A common choice is a univariate Gaussian distribution:

$$P(N \mid \theta_j) = \frac{1}{\sigma \sqrt{2\pi}} e^{-(N - \mu_j)^2 / (2\sigma^2)} \qquad \text{(B.33)}$$

where μ_j and σ are the parameters of the distribution for $j = 1, 2$. We have made an implicit assumption that the variance is the same under each stimulus. This is called a *homoscedastic assumption*. Parameters, μ_j and σ can be estimated by taking the empirical mean and standard deviation from spike count data. To estimate σ, you can use

$$\sigma = \frac{\sigma_1 + \sigma_2}{2}$$

where $\sigma_j^2 = \mathrm{var}(N \mid \theta_j)$ is the sample variance under stimulus $j = 1, 2$. We should not use the unconditioned sample variance, $\mathrm{var}(N)$, to estimate σ^2 because taking the variance across different stimuli would lead to an artificially increased estimate of the variance (why?).

Equation (B.33) is a *statistical model* of neurons' spike counts. This is fundamentally different from the other models in this book, which are derived from the actual biophysics of the neurons themselves. Statistical models allow us to abstract away from the physical neurons. This Gaussian model is obviously a rough approximation because spike counts must be positive integers, but this approximation will still allow us to derive useful decoding methods.

Under the model in equation (B.33), we can simplify (B.32) to

$$\text{Guess } \theta_1 \text{ whenever } |N - \mu_1| < |N - \mu_2|. \qquad \text{(B.34)}$$

In other words, we should guess θ_1 if N is closer to μ_1 and guess θ_2 if N is closer to μ_2. We can estimate μ_1 and μ_2 by averaging the spike counts under each condition. We can visualize the decision algorithm in equation (B.34) by drawing a vertical line at the midpoint:

$$z = \frac{\mu_1 + \mu_2}{2}$$

between the two means (see the black dotted line in figure B.10a). Then our guess is determined by which side of the vertical line our spike count lies on. Whenever $\mu_1 > \mu_2$, we can write the decision algorithm in equation (B.34) as

$$\text{Guess } \theta_1 \text{ whenever } N > z. \qquad \text{(B.35)}$$

If μ_2 were larger, then we would instead guess θ_2 whenever $N > z$. When we apply this algorithm to the data in figure B.10A, we guess correctly on 78 percent of the trials.

The decision algorithm will guess correctly more often (*i.e.*, the two angles are easier to distinguish) when the mean spike counts are farther apart and/or the spike count variance is smaller. To this end, we can define the *signal-to-noise ratio (SNR)*:

$$SNR = \frac{(\mu_2 - \mu_1)^2}{\sigma^2}.$$

When *SNR* is large, it is easier to decode the orientations from the spike counts. When *SNR* is small, it is more difficult.

Monkeys are perfectly capable of discriminating between a 120-degree and a 150-degree orientation, but we were only able to guess correctly on 78 percent of the trials. What gives? One problem is that we only used the spike train of one neuron, but the monkey's visual cortex has millions of neurons. Figure B.10b shows the spike count histograms for a second neuron that was recorded simultaneously with the neuron in figure B.10a. Applying the same decision algorithm to this neuron leads us to guess correctly on 66 percent of the trials. And figure B.10c shows a scatterplot of the spike counts of both neurons. This plot potentially contains more information than figures B.10a and B.10b combined because it shows how the two neurons' spike counts change together. How can we use this extra information to improve our decoding accuracy? How can we use the information from a large population of simultaneously recorded neurons to decode the stimulus? Answering these questions has driven a large body of research in *neural population coding*, which is the study of how information is encoded in populations of neurons' spike trains and how it can be decoded.

Decoding a population of neurons. Let's now consider a heterogeneous population of K neurons. Let $N = [N_1 \, N_2 \, \ldots N_K]^T$ be a $K \times 1$ multivariate random variable representing the spike counts of all K neurons. Generalize the Gaussian model given earlier to a *multivariate Gaussian* model with probability density function

$$P(N \mid \theta_j) = \frac{1}{\sqrt{2\pi |\Sigma|}} \exp\left(-\frac{1}{2}(N - \mu^j)^T \Sigma^{-1} (N - \mu^j) \right) \tag{B.36}$$

where $|\Sigma|$ is the determinant of Σ. Here, $\mu^j = E[N \mid \theta_j]$ is the expectation of N under stimulus $j = 1, 2$. Hence, μ_k^j is the mean spike count of neuron k under stimulus j. The covariance matrix, Σ, quantifies the covariance between spike counts. Specifically, Σ is a $K \times K$ matrix with entries defined by

$$\Sigma_{k,k'} = \text{cov}(N_k, N_{k'})$$

for $k, k' = 1, \ldots, K$. The covariance matrix, Σ, is symmetric and positive semi-definite. In other words, it must satisfy $\Sigma_{k,k'} = \Sigma_{k',k}$ (*i.e.*, $\Sigma^T = \Sigma$), all of its eigenvalues are real, and none of its eigenvalues are negative. Note that we again make a homoscedastic assumption; that is, we assume that the covariance matrix is Σ under both stimuli. We can estimate Σ by averaging the covariance matrix under the two stimuli:

$$\Sigma = \frac{\Sigma_1 + \Sigma_2}{2}$$

where Σ_j is the covariance matrix under stimulus j. In NumPy, the covariance matrix can be estimated from data by doing

```
Sigma1=np.cov(SpikeCounts[:,0,:])
Sigma2=np.cov(SpikeCounts[:,1,:])
Sigma=(Sigma1+Sigma2)/2
```

where `SpikeCounts[i,j,k]` contains the spike count of neuron k on trial i of stimulus j. This code averages the covariance matrices under the two conditions. The spike count covariance, $cov(N_k, N_{k'})$, quantifies the amount of trial-to-trial variability shared by the two neurons. This shared variability is more commonly quantified by the *spike count correlation*:

$$c = c_{k,k'} = \frac{cov(N_k, N_{k'})}{\sqrt{var(N_k)var(N_{k'})}} = \frac{\Sigma_{k,k'}}{\sqrt{\Sigma_{k,k}\Sigma_{k',k'}}},$$

also known as the *Pearson correlation coefficient*. We will write c when it's not necessary to identify which neurons we are talking about and $c_{k,k'}$ when it is necessary. Note that $c_{k,k} = 1$ and $c_{k,k'} \in [0, 1]$ for all k and k'. The matrix of correlation coefficients can be found using `np.corrcoef` in place of `np.cov` in the previous code snippet.

The correlation measures the *proportion* of shared variability between N_k and $N_{k'}$. For Gaussian distributions, the value of c determines how skewed we should expect the clouds of points in figure B.10c to be. If $c = 0$, the spike counts are uncorrelated and the clouds should be approximately circular. If $c > 0$, then the clouds are skewed along a diagonal line with positive slope and larger value of c correspond to more tightly skewed clouds. If $c = 1$, then all of the points would lie on a line with positive slope. If $c < 0$, then they are skewed along a diagonal line with negative slope. The data in figure B.10c has a sample correlation coefficient of $c = 0.15$ under stimulus θ_1 and 0.16 under θ_2. Figure B.11 shows some artificially generated spike counts with different correlation values.

Assuming again that both stimuli are equally likely, $P(\theta_1) = P(\theta_2) = 0.5$, the optimal decision algorithm (B.32) simplifies to (McLachlan 2005; Dayan and Abbott 2001)

$$w = \Sigma^{-1}(\mu_1 - \mu_2)$$

$$z = w \cdot (\mu_1 + \mu_2)/2 \tag{B.37}$$

Guess θ_1 if $w \cdot N > z$ and guess θ_2 if $w \cdot N < z$.

This decision algorithm is called *Fisher's linear discriminant analysis (Fisher's LDA)*.

To visualize Fisher's LDA, note that the set of all N satisfying $w \cdot N = z$ defines a *hyperplane* orthogonal to w that contains the point $E[N] = \mu = (\mu_1 + \mu_2)/2$. A hyperplane is a generalization of a plane. When $K = 2$, a hyperplane is a line. When $K = 3$, it is a plane. For any K, a hyperplane is a $(K-1)$-dimensional surface that cuts the space into two halves. One half contains points satisfying $w \cdot N > z$, and the other half contains points satisfying $w \cdot N < z$. The decision algorithm (B.37) defines the optimal hyperplane for classifying stimuli under the multivariate Gaussian model in equation (B.36). The dashed line in figure B.10c shows the optimal hyperplane for the data in the figure. If the spike counts of the two neurons lie above the dashed line, we should guess θ_1. If the spike counts are below the line, we should guess θ_2.

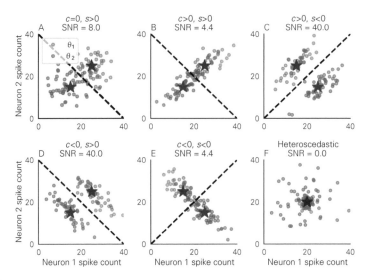

Figure B.11
Synthetic spike counts to visualize the effects of correlations on decoding accuracy. (A–E) Random spike
counts sampled from homoscedastic bivariate Gaussian distributions (equation (B.36) and (B.38) with $K = 2$) with
various noise and stimulus correlations (c and s). (F) Same, but with heteroscedastic distributions. SNRs were
computed from equation (B.39). See the second code cell of `PopulationCoding.ipynb` for the code to
produce this figure.

We can visualize the impact of correlations on decoding by generating synthetic data.
Consider the model equation (B.36) with $K = 2$. The covariance matrix can be written as

$$\Sigma = \begin{bmatrix} \sigma^2 & c\sigma^2 \\ c\sigma^2 & \sigma^2 \end{bmatrix} \tag{B.38}$$

where c is the correlation coefficient between the spike counts of neuron 1 and neuron 2.
Figure B.11a shows points drawn randomly from this model with $c = 0$. The more the two
point clouds overlap, the more difficult it is to discriminate the two stimuli, and the more
errors we will make with our decision algorithm. Figure B.11b shows the model with the
same parameters except $c > 0$. Positive correlations stretch the distributions along an axis
with positive slope, causing the distributions to overlap more and therefore making it more
difficult to discriminate between the stimuli. This result might lead you to conclude that
positive correlations are bad for coding, but consider the data in figure B.11c, which were
drawn from the same model with $c > 0$ except that the mean spike counts are different. In
this case, correlations decrease the overlap between the point clouds and therefore improve
coding.

How can we understand the salient difference between figure B.11a–e? The *signal correlation* between two spike counts is the correlation coefficient between the neurons' *mean*
firing rates as the stimulus varies (*i.e.*, the correlation between $E[N_1 \mid \theta]$ and $E[N_2 \mid \theta]$ as θ
varies). Positive signal correlations imply that the two neurons change their mean firing rates
in a similar way when the stimulus changes (when one increases, the other also increases).
Negative signal correlations imply that they change their mean firing rates oppositely (when
one increases, the other decreases). When we are considering only two stimuli, θ_1 and θ_2,

as in the case here, the stimulus correlation is either 1, -1, or 0. Specifically,

$$s_c = \text{sign} \left(\frac{E[N_2 \mid \theta_2] - E[N_2 \mid \theta_1]}{E[N_1 \mid \theta_2] - E[N_1 \mid \theta_1]} \right)$$

$$= \text{sign} \left(\frac{\mu_{2,2} - \mu_{2,1}}{\mu_{1,2} - \mu_{1,1}} \right)$$

where μ_{kj} is the mean spike count of neuron k under stimulus θ_j and $\text{sign}(x) = x/|x|$ returns 1 when $x > 0$ and -1 when $x < 0$. If you draw a line connecting the two means in figure B.10c (*i.e.*, from one star to the other), then $s_c = 1$ if the slope is positive and $s_c = -1$ if the slope is negative. The regular correlation coefficient, c, sometimes called the *noise correlation* to distinguish it from the signal correlation. The idea is that c measures the similarity between the noise in the spike counts (*i.e.*, the variability around the means) while s_c measures the similarity between changes in the signals (*i.e.*, the means).

You can see from the examples in figure B.11a–e that correlations improve coding whenever the signal correlation, s_c, has the opposite sign as the noise correlation, c, because the distributions overlap less in this case. Conversely, when s_c and c have the same sign, correlations make coding worse because the distributions overlap more.

While figure B.11a–e provides nice intuition, it is restricted to the case of $K = 2$ neurons, and we should also seek a more precise quantification of discriminability. To achieve this, we can generalize the SNR here to multiple neurons. To derive the SNR, first define

$$u = \boldsymbol{w} \cdot N.$$

Because linear operations preserve Gaussianity, u obeys a univariate Gaussian distribution whenever N obeys a multivariate Gaussian distribution. Therefore, equation (B.37) corresponds to applying a univariate decision algorithm (like the one in equation (B.34)) to $u = \boldsymbol{w} \cdot N$. Hence, we can define the SNR of equation (B.37) in terms of the mean and variance of u:

$$SNR = (\boldsymbol{\mu}_2 - \boldsymbol{\mu}_1)^T \Sigma^{-1} (\boldsymbol{\mu}_2 - \boldsymbol{\mu}_1). \tag{B.39}$$

This expression for the SNR is applicable for any number of neurons, K. The SNR values are given in figure B.11 and confirm our intuitions.

Note that the optimality of equation (B.37) depends on our homoscedastic assumption ($\Sigma_1 = \Sigma_2 = \Sigma$). Figure B.11f shows an example under a heteroscedastic model in which the variance is larger under stimulus 2. The means are the same ($\boldsymbol{\mu}_1 = \boldsymbol{\mu}_2$), so equation (B.37) is ineffective and the SNR given by equation (B.39) is zero. However, the variances are different under the two distributions, which allows us to do better than chance when decoding the stimuli. In particular, we can guess θ_2 (blue) whenever the spike counts are farther from their mean, and θ_1 (red) when they are closer. This example demonstrates that the decision algorithm in equation (B.37) is not optimal for heteroscedastic models, but optimal decision algorithms for heteroscedastic models do not always have simple, linear decision boundaries like the hyperplane defined by equation (B.37).

The interpretation of equation (B.37) as performing univariate discrimination on u gives us another way to visualize equation (B.37) and the SNR from equation (B.39). In particular, we can plot histograms of u under each condition and draw the thresholds at z (much like

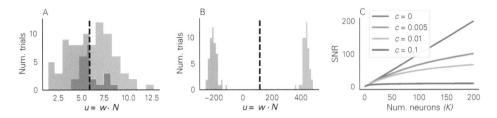

Figure B.12
Population decoding. (A) The histogram of $u = \mathbf{w} \cdot \mathbf{N}$ under two stimulus conditions computed from the $K = 2$ visual cortical neurons in figure B.10. (B) Same as (A), but for $K = 112$ neurons. (C) The SNR as a function of the number of neurons (equation (B.40)) for different values of c under a model with homogeneous spike count statistics (Zohary, Shadlen, and Newsome 1994). This model ignores heterogeneous spike count statistics and can thereby lead to an incorrect conclusion that correlations are always detrimental to coding in large populations. See the first code cell of `PopulationCoding.ipynb` for the code to produce this figure.

the histograms of N in figures B.10). Figure B.12a shows this visualization applied to the $K = 2$ data from figure B.10. The histograms are slightly better separated than they are in figures B.10a,b. Figure B.12b shows the same visualization applied to $K = 112$ neurons from the same experiments. The histograms are extremely well separated, implying that the two angles can easily be distinguished with enough neurons.

A historical note.   Early theoretical work on the impact of correlations in neural coding assumed a homogeneous population of neurons (*i.e.*, all neurons have the same mean spike count under each stimulus and all pairs of neurons have the same correlation). In other words,

$$\boldsymbol{\mu}_j = E[\mathbf{N} \mid \theta_j] = \left[\mu_j \, \mu_j \ldots \mu_j\right]^T$$

for $j = 1, 2$ and

$$\Sigma_{k,k'} = \begin{cases} c\sigma^2 & k \neq k' \\ \sigma^2 & k = k'. \end{cases}$$

Under these assumptions,

$$SNR = \frac{SNR_1}{c + (1 - c)/K} \tag{B.40}$$

where $SNR_1 = (\mu_2 - \mu_1)^2/\sigma^2$ is the SNR for $K = 1$ neuron. Figure B.12c shows the SNR under this homogeneous model for various values of c. When $c = 0$, the SNR grows without bound. However, even small values of $c > 0$ have a large impact on the SNR when K is large. Note that we cannot have $c < 0$ at large K because this would violate the requirement that Σ is positive semi-definite. It was concluded that positive correlations are detrimental to coding, even when they are weak. This result can be understood as follows: When $c = 0$, the noise in the spike counts is independent, so it can be averaged out by summing all K spike counts, so the SNR grows without bound for increasing K. When $c > 0$, there is some shared noise that cannot be averaged out, so the SNR is bounded even at large K.

This argument that correlations are detrimental to coding was famously laid out by Ehud Zohary, Micheal Shadlen, and Willian (Bill) Newsome in 1994 (Zohary, Shadlen, and Newsome 1994). The resulting conclusion that *correlated spike counts are bad for coding* remained a widely repeated mantra in computational neuroscience for a decade before the subtleties about noise versus signal correlations (described previously) became more widely known and accepted (Averbeck and Lee 2006; Averbeck, Latham, and Pouget 2006). The story of the impact of spike train correlations on neural coding has developed further from there (Moreno-Bote et al. 2014; Panzeri et al. 2022). Some researchers have argued that this picture of neural coding, in which each neuron encodes some stimulus feature in its firing rate, is oversimplified, counterproductive, or just wrong (Brette 2019).

Fisher information and its relationship to SNR. Animals are often trained to distinguish among similar orientations ($d\theta = \theta_2 - \theta_1$ small) and their discrimination errors are compared to neural recordings and stimulus properties. These experiments are often modeled using the same model as before, in the limit of a small difference between stimuli ($d\theta = \theta_2 - \theta_1 \to 0$). If

$$\boldsymbol{\mu}(\theta) = E[N \mid \theta]$$

is a continuous function of θ, then

$$\lim_{\theta_2 \to \theta_1} SNR = 0.$$

since $\boldsymbol{\mu}_2 \to \boldsymbol{\mu}_1$ as $\theta_2 \to \theta_1$. But we can normalize by $d\theta^2 = (\theta_2 - \theta_1)^2$ to define the *Fisher information*:

$$F(\theta_1) = \lim_{\theta_2 \to \theta_1} \frac{SNR}{(\theta_2 - \theta_1)^2} = \boldsymbol{\mu}'(\theta_1)^T \Sigma^{-1} \boldsymbol{\mu}'(\theta_1)$$

where $\boldsymbol{\mu}'(\theta)$ is the K-dimensional vector of derivatives of $\boldsymbol{\mu}(\theta)$. Mathematical properties of the Fisher information and its dependence on experimental parameters can produce experimentally testable predictions that can be compared to behaviour and/or neural recordings from animals that are trained to discriminate among similar orientations. Many of the properties of SNR that we describe here were originally described in the literature in terms of the Fisher information instead of the SNR, but the two quantities are analogous.

Exercise B.4.1. The homoscedastic assumption makes the decision algorithm simpler when using a Gaussian model. However, the Gaussian model in equation (B.33) and the homoscedastic assumption are somewhat unrealistic for spike counts. Spike counts for a single neuron might be better fit by a Poisson model of the form

$$P(N \mid \theta_j) = \frac{\mu_j^N}{N!} e^{-\mu_j}.$$

Here, μ_j is both the mean *and* the variance of the distribution, so the model is heteroscedastic despite having only one parameter. Derive an optimal decision algorithm for this Poisson model starting from equation (B.32). How does the heteroscedasticity change the decision boundary? Apply this decision algorithm to the data from figure 2.3b.

Exercise B.4.2. Reproduce figure B.12b using a subset of the neurons. See how the discriminability changes as you include a larger or smaller number of neurons.

Exercise B.4.3. **Do correlations help or hurt coding in a real neural population?** Using the data from $K = 112$ neurons from figure B.12b, compute the SNR. Then create a new

Σ by setting all values except for the diagonals to zero (you can use `SigmaNew=np.diag(np.diag(Sigma))`). Compute the SNR from this new Σ that ignores correlations. Which SNR is larger?

Important note: Since Σ is a large, ill-conditioned matrix, you should use the pseudo-inverse (`np.linalg.pinv`) in place of the actual inverse to compute the SNR.

B.5   Derivations and Alternative Formulations of Rate Network Models

In section 3.3 of chapter 3, we introduced dynamical rate network models, but we did not show exactly how they are derived from spiking networks. In this section, we give a derivation as well as alternative formulations of rate models.

Let's begin by considering the equation for the synaptic input from population b to population a in a recurrent spiking network model in equation (3.4):

$$\tau_b \frac{d\mathbf{I}^{ab}}{dt} = -\mathbf{I}^{ab} + J^{ab} \mathbf{S}^b, \quad a = e, i, \ b = e, i, x.$$

There are two sources of randomness in the network: randomness in the entries of J^{ab} and randomness in the spike times in $\mathbf{S}^b(t)$. Fortunately, since this is a linear system of ODEs, we can take expectations of both sides of the equation. Linear ODEs interact well with expectations: the expectation of the solution is given by solving an ODE for the expectations. Note that each term of the matrix-vector product is a sum of the form

$$\left[J^{ab} \mathbf{S}^b \right]_j = \sum_{k=1}^{N_b} J_{jk}^{ab} S_k^b$$

Therefore, if we take the expectation over both sources of randomness, we get

$$\tau_b \frac{dI_{ab}}{dt} = -I_{ab} + w_{ab} r_b(t) \tag{B.41}$$

where $I_{ab}(t) = E[I_j^{ab}(t)]$ does not depend on j, and r_b is the mean instantaneous firing rate of neurons in population b. The mean-field synaptic weight is defined by $w_{ab} = N_b E[J_{jk}^{ab}]$. For the random connectivity model considered in this text (see equation (3.5) in chapter 3), this gives $w_{ab} = N_b p_{ab} j_{ab}$, but other connectivity models can be used instead. Equation (B.41) is exact for external presynaptic inputs ($b = x$). For recurrent inputs ($b = e, i$), equation (B.41) is technically an approximation because its derivation assumed that J_{jk}^{ab} and $S_k^b(t)$ are uncorrelated, which is not a valid assumption. Nevertheless, it is an accurate approximation in practice. Taken together, equation (B.41) represents a system of six ODEs, one for each value of $a = e, i$ and $b = e, i, x$.

Equation (B.41) is not useful by itself because we do not know r_e and r_i. They are determined by the complicated dynamics of the network. Deriving a mean-field equation for r_e and r_i is not so simple due to the highly nonlinear dynamics of the EIF neurons that determine the mapping from inputs to firing rates. Instead, we can use a mean-field approximation given by

$$\tau_{r,e} \frac{dr_e}{dt} = -r_e + f_e(I_{ee} + I_{ei} + I_{ex}) \tag{B.42}$$

$$\tau_{r,i}\frac{dr_i}{dt} = -r_i + f_i(I_{ie} + I_{ii} + I_{ix})$$

$$\tau_b\frac{dI_{ab}}{dt} = -I_{ab} + w_{ab}r_b$$

(B.43)

where $\tau_{r,a}$ is a time constant modeling how quickly firing rates settle to their fixed points, and f_e is an f-I curve for population $a = e, i$. Typically, we would take $\tau_{r,e} = \tau_{r,i} = \tau_r$ and $f_e = f_i = f$ if neurons in both populations are the same or similar. We might also take $\tau_r \approx \tau_m$ since firing rates dynamics are determined by membrane potential dynamics. Equation (B.43) gives a system of eight ODEs that can be solved to approximate the dynamics of a recurrent network. Fixed points of equation (B.43) satisfy equation (3.8), so the dynamical mean-field theory is consistent with the stationary mean-field in equation (3.8).

Equation (B.43) can easily be extended to networks with more populations. It can also be modified to derive mean-field equations for feedforward networks, but this is rarely done because the dynamics are usually uninteresting. However, equation (B.43) is rarely used in practice because it is unnecessarily high-dimensional, which can make it more difficult to study. Ideally, we would have one ODE for each population in the recurrent network; that is, we would have a system of two ODEs for the network considered here. Systems of two ODEs are especially nice because they are easier to study using the dynamical systems methods from appendix A.8.

We next discuss three approaches to obtaining dynamical mean-field models with one ODE for each population.

Reducing the dimension of the mean-field equations by ignoring synaptic dynamics. The first approach ignores synaptic dynamics by omitting the last equations for $I_{ab}(t)$ and replacing them with

$$I_{ab} = w_{ab}r_b$$

in the first two equations. This is called a **quasi-steady state approximation** since it replaces I_{ab} with the value that its fixed point ("steady state") would take if r_b were fixed in time (see the derivations in section 2.3 of chapter 2). This replacement would be mathematically justified if the synaptic time constants were much smaller than the rate time constants ($\tau_e, \tau_i \ll \tau_r$) since $I_{ab}(t)$ would converge to its fixed point much faster than $r_a(t)$. In reality, the τ values are similar in magnitude, so this assumption is not justified. In any case, the substitution produces a much simpler set of equations:

$$\tau_{r,e}\frac{dr_e}{dt} = -r_e + f_e(w_{ee}r_e + w_{ei}r_i + X_e)$$

$$\tau_{r,i}\frac{dr_i}{dt} = -r_i + f_i(w_{ie}r_e + w_{ii}r_i + X_i)$$

(B.44)

where $X_a = w_{ax}r_x$ is the external input. However, equation (B.44) completely ignores synaptic dynamics and the effects of synaptic timescales (τ_e and τ_i), which prevents it from capturing some phenomena. For example, equation (B.44) would not capture the oscillations studied in section 3.3 in chapter 3 since they were caused by a mismatch between the timescales of excitatory and inhibitory synapses (τ_e and τ_i). Of course, we only need to replace $\tau_{r,e}$ and $\tau_{r,i}$ in equation (B.44) by τ_e and τ_i to get equation (3.10). Hence, the only difference between equations (3.10) and (B.44) is the interpretation of the time constants. We return to this point toward the end of this section.

Reducing the dimension of the mean-field equations by ignoring rate dynamics. The second approach to reducing the dimension of equation (B.43) ignores rate dynamics by omitting the first equations for r_e and r_i and instead makes the following substitution:

$$r_a = f(I_{ae} + I_{ai} + I_{ax})$$

in the first two equations. This replacement would be mathematically justified if the firing rates evolved much more quickly than the synaptic dynamics ($\tau_r \ll \tau_e, \tau_i$). We can also write the external input explicitly as a time series $I_{ax}(t) = X_a(t)$. Together, this yields a system of four ODEs:

$$\tau_b \frac{dI_{ab}}{dt} = -I_{ab} + w_{ab}f(I_{be} + I_{bi} + X_b), \quad a,b = e,i. \tag{B.45}$$

This system of equations accounts for synaptic timescales (so it can be used to describe the oscillations in section 3.3), but it is four-dimensional, whereas we would prefer a two-dimensional system for two populations (e and i).

Equation (B.45) can be reduced to a system of two equations by taking $\tau_e = \tau_i = \tau$ and then modeling the total synaptic inputs, $I_e = I_{ee} + I_{ei}$ and $I_i = I_{ie} + I_{ii}$. These two simplifications allow us to write the synaptic inputs as a system of two equations:

$$\tau \frac{dI_e}{dt} = -I_e + w_{ee}f(I_e + X_e) + w_{ei}f(I_i + X_i)$$
$$\tau \frac{dI_i}{dt} = -I_i + w_{ie}f(I_e + X_e) + w_{ii}f(I_i + X_i) \tag{B.46}$$

which is more often written as

$$\tau \frac{dI_e}{dt} = -I_e + w_{ee}r_e + w_{ei}r_i$$
$$\tau \frac{dI_i}{dt} = -I_i + w_{ie}r_e + w_{ii}r_i$$
$$r_e = f(I_e + X_e)$$
$$r_i = f(I_i + X_i). \tag{B.47}$$

Even though equation (B.47) is written as four equations, it is still a system of two ODEs, and it is completely equivalent to equation (B.46). Generalizing this approach to an arbitrary number of populations gives equation (3.13) from section 3.3.

Equation (B.47) is a two-dimensional system, but it does not account for two synaptic timescales because we had to take $\tau_e = \tau_i = \tau$ to derive it. Therefore, like equation (B.44), it cannot capture the oscillations studied in section 3.3. Sometimes two synaptic timescales, τ_e and τ_i, are used in place of τ in equation (B.47). However, this approach is not mathematically justified because it implicitly assumes that synaptic time constants depend on the *postsynaptic* cell type, whereas they should depend on the *presynaptic cell type*. For example, this approach would assume that the timescale of $I_e = I_{ee} + I_{ei}$ is τ_e, whereas the timescale of I_{ei} is τ_i.

Reducing the dimension of the mean-field equations by modeling filtered firing rates. The third approach to reducing the dimension of equation (B.43) requires a little trick that is not obvious at first but gives a nicer result in the end. The trick is to define new quantities, $y_e(t)$ and $y_i(t)$, satisfying

$$\tau_a \frac{dy_a}{dt} = -y_a + r_a.$$

These are just the firing rates passed through a low-pass filter–that is, they are the firing rates convolved with an exponential kernel (see appendix A.4). Now, note that the synaptic inputs from equation (B.43) can be written in terms of $y_a(t)$ using

$$I_{ab}(t) = w_{ab} y_b(t).$$

Now, we make the same substitution:

$$r_a = f(I_{ae} + I_{ai} + X_a)$$

for the rates that we made for the previous approach. Combining these three equations gives

$$\tau_e \frac{dy_e}{dt} = -y_e + f(w_{ee} y_e + w_{ei} y_i + X_e)$$

$$\tau_i \frac{dy_i}{dt} = -y_i + f(w_{ie} y_e + w_{ii} y_i + X_i).$$

(B.48)

Equation (B.48) is a system of two equations that captures both synaptic timescales separately, which was our goal. The only caveat is that $y_a(t)$ represents a, filtered firing rate instead of the firing rate itself. However, the dynamics should be similar. For this reason, equation (B.48) is often used with y_a replaced by r_a to get equation (3.10) from section 3.2. Generalizing to an arbitrary number of populations gives equation (3.12) from section 3.3.

Equations (B.44) and (B.48) have the same mathematical form but different interpretations. Most notably, the time constants in equation (B.44) correspond to the timescale at which *firing rates* evolve, while the time constants in equation (B.48) correspond to the timescale at which *synaptic currents* evolve. This difference arises because the derivation of equation (B.44) ignored synaptic dynamics with the substitution $I_{ab} = w_{ab} r_b$, while the derivation of equation (B.48) ignored rate dynamics by making the substitution $r_a = f(I_{ae} + I_{ai} + X_a)$. Neither of these substitutions is justified. Instead of using one interpretation or the other, we can interpret τ_e and τ_i as the *combined* timescales of the synapses and firing rate dynamics. In particular, populations with slower synapses *or* neuron dynamics should have larger time constants. This is the interpretation used in sections 3.2 and 3.3. This interpretation is not precise and does not tell you exactly how to choose the actual values of the time constants, but it is still useful. If you need to be more precise in accounting for the various time constants, then you should use the system of eight ODEs described by equations (B.41) and (B.43), but this might not provide much more insight or simplicity than a spiking model itself.

B.6 Hopfield Networks

John Hopfield is an Amer*can physic*st wh* is
famo*s f*r devel*ping a c*mput*tional m*del

```
of pattern completion now kn*wn as th*
Hopfield model. Pat*ern compl*tion o*curs
wh*n y*u obs*rve a part*ally obsc*red stim*lus
and y*ur bra*n fi*ls in the mis*i*g d*tails. For
exa*ple, you are pro*ab*y ab*e to re*d th*s
p*ragr*ph ev*n tho*gh m*ny let*ers are mis*i*g.
```

In case you are unable to parse the paragraph above, here is an unedited version:

John Hopfield is an American physicist who is famous for developing a computational model of *pattern completion*, now known as the *Hopfield model* (Hopfield 1982). Pattern completion occurs when you observe a partially obscured stimulus and your brain fills in the missing details. For example, you are probably able to read this paragraph even though many letters are missing.

As another example of pattern completion, consider the image in figure B.13a. Despite the fact that some pixels are corrupted, you can tell that the image is a handwritten 3 (figure B.13b) and you can probably repair the corrupted pixels. Pattern completion is important and common because stimuli that you observe on a daily basis are often partially obscured. Pattern completion is an example of *associative memory*, in which stimulus features that commonly appeared together in the past are associated with one another so that the presentation of one stimulus feature conjures memories of the others. For example, you may associate certain smells with your home. Those smells will conjure memories of your home even when you come across them in different settings.

The basic idea behind Hopfield networks is to build a network with a stable fixed point associated with each stimulus that needs to be remembered. These fixed points are called *attractor states* or *attractors*. When a corrupted stimulus is presented, the network converges to the nearest attractor, which should be the uncorrupted stimulus. The neurons that are activated by a particular stimulus are sometimes called a *neural assembly*. In a Hopfield network, if part of an assembly is activated by a partially obscured stimulus, then they should activate the other neurons in the assembly to complete the stimulus, hence the name "pattern completion." For this to occur, neurons that are frequently activated together by a particular

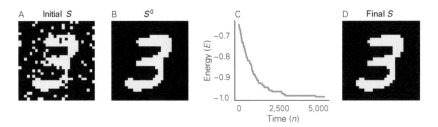

Figure B.13
Pattern completion in a Hopfield network. (A) A corrupted image of a handwritten digit. The image was corrupted by flipping the sign of eighty randomly chosen pixels. This pattern was used as the initial condition in a Hopfield network with $N = 28 * 28 = 784$ neurons. (B) The original uncorrupted image. This pattern was trained into the Hopfield network by setting the weights according to equation (B.51). (C) The energy of the resulting Hopfield network during a simulation. (D) The final state of the Hopfield network matches the trained attractor state. See `HopfieldNetwork.ipynb` for the code to produce this figure.

stimulus should be connected more strongly with positive weights, which is a key component of Hebbian plasticity (see section 4.1 of chapter 4). Moreover, since each stimulus should be associated with only one attractor state, the activation of one neural assembly should suppress the others, which is a form of suppression or competition (see section 3.3).

Now let's define the network more precisely. Hopfield networks are recurrent binary neural networks (see section B.2.3) defined in discrete time, $n = 1, 2, \ldots$ instead of continuous time. In place of spike trains, Hopfield networks have an N-dimensional vector, $S(n)$. At each point in time, each neuron is in one of two states:

$$S_j(n) = 1 \text{ or } S_j(n) = -1$$

where 1 corresponds to the spiking or active state, and -1 corresponds to the silent or nonspiking state. The N-dimensional vector of inputs is defined by

$$I(n) = W S(n)$$

where W is an $N \times N$ connectivity matrix. At each time step, one neuron, j, is chosen at random to update. The randomly chosen neuron is updated according to the following rule:

$$S_j(n+1) = \begin{cases} 1 & I_j(n) \geq 0 \\ -1 & I_j(n) < 0. \end{cases} \tag{B.49}$$

This practice of choosing a random neuron to update is called a *stochastic update* scheme. Since only neuron j is updated on a particular time step, you need to compute $I_j(n)$ just on that time step, not all of I. In summary, the following steps define how to run a Hopfield network:

1. Choose a random neuron, j.
2. Compute $I_j = \sum_k W_{jk} S_k$.
3. Update S_j using equation (B.49).
4. Repeat. This process is repeated until $S(n)$ converges toward a fixed point.

A major advantage of the Hopfield model is that, under certain conditions, we can understand fixed points and their stability very well. To this end, define the **energy** the network in state, S, as

$$E = -\sum_{j,k=1}^{N} W_{jk} S_j S_k = -S^T W S. \tag{B.50}$$

The term "energy" is often used for a quantity that is either conserved or strictly decreasing in a model. Indeed, the following theorem is central to the study of Hopfield networks:

Theorem. If network connectivity is symmetric ($W_{jk} = W_{kj}$) and there are no self-connections ($W_{kk} = 0$), then E decreases over time and $S(n)$ converges to fixed points, S^0, that are local minima of E.

The goal, then, is to learn a matrix, W, for which the target stimuli are local minima of E. For example, if we want to perform pattern completion on digits like the one in figure B.13a, then we can define a Hopfield network with $N = 28 * 28 = 784$ neurons, each representing one pixel. If each of the ten digits is a local minimum of E, then they will be attractor states for the network. Then, if we start with a corrupted digit as an initial condition, the network will converge to the nearest attractor state, which is likely to be the original digit.

How can we choose W to promote particular attractor states? Suppose that S^0 is a vector that we want to be an attractor state. Then we want E to be smaller (*i.e.*, more negative) for that particular state. To achieve this, we would want to set $W_{jk} > 0$ whenever $S_j^0 = S_k^0 = 1$. Note that we *also* want to set $W_{jk} > 0$ whenever $S_j^0 = S_k^0 = -1$ (why?). On the other hand, we want to set $W_{jk} < 0$ whenever $S_j^0 \neq S_k^0$. Recall that we must also have $W_{jk} = W_{kj}$ and $W_{kk} = 0$. Therefore, if there is just one desired attractor state, we can set

$$W_{jk} = \begin{cases} cS_j^0 S_k^0 & j \neq k \\ 0 & j = k. \end{cases} \tag{B.51}$$

where $c > 0$ can be any constant. The choice $c = 1/N^2$ allows equation (B.50) to be interpreted as an average. This can be implemented efficiently in NumPy as

```
W=(1/N**2)*(np.outer(S0,S0)-np.diag(S0))
```

Here, `np.outer(S0,S0)` returns the outer product of `S0` with itself, and `diag(S0)` returns a diagonal matrix with `S0` along the diagonal. Figure B.13c shows the energy of a Hopfield network simulation with initial condition given by the pattern in figure B.13a, where W was chosen using equation (B.51) with S^0 being the pattern in figure B.13b. Figure B.13d shows the final state of the Hopfield network.

Note that if S^0 is an attractor, then so is $-S^0$ since they have the same energy, so equation (B.51) produces two attractors. However, if the initial condition, $S(0)$, is closer to S^0, then we should expect $S(n)$ to converge toward S^0 instead of $-S^0$.

Of course, we would like to be able to train more than one attractor state. Now, suppose that there are M attractors, $\{S^m\}_{m=1}^M$. We would like the Hopfield network to converge to the attractor that is closest to the initial condition. To achieve this, we can just average the Ws that we would obtain from equation (B.51):

$$W_{jk} = \begin{cases} \frac{1}{MN^2} \sum_{m=1}^M S_j^m S_k^m & j \neq k \\ 0 & j = k. \end{cases} \tag{B.52}$$

If we imagine that the stimuli, S^m, are presented to the network sequentially during a learning phase, then equation (B.52) can be written as a Hebbian plasticity rule:

$$W_{jk} = W_{jk} + \eta S_j S_k$$

which is applied sequentially for $S = S^1, S^2, \ldots, S^M$ at all indices $j \neq k$.

Note that $S_j^m S_k^m = 1$ when $S_j^m = S_k^m$ and $S_j^m S_k^m = -1$ when $S_j^m \neq S_k^m$. Hence, this update rule increases W_{jk} whenever S_j^m and S_k^m tend to be active together in the training stimuli, and it decreases W_{jk} when they do not tend to be active together. In summary, this rule enforces a form of Hebbian plasticity and suppression or competition: Neurons that fire together wire together in the sense that they form positive connections, while neurons that do not fire together suppress each other through negative connections.

There is no guarantee that every S^m will be an attractor state after training, but if all the S^m are sufficiently far away from each other and m is not too large, then it is likely. If some S^m values are close to each other, then nearby S^m can start to interfere with each other's stability.

Numerous generalizations of Hopfield networks have been developed, but the overall concept is often the same: Memories are stored as attractors in a recurrent network and convergence to these attractors is quantified by the minimization of an energy function. Models of this type are sometimes called *energy-based models*. Hopfield networks and other energy-based model are a vast abstraction away from biological neural circuits, and they only solve a relatively simple task. These factors might lead you to question their relevance and usefulness, but they are more useful than they appear at first.

One useful property of Hopfield networks and many other energy-based models is that the learning dynamics can often be understood mathematically. For example, under various conditions, you can derive how a network's memory capacity (*i.e.*, the maximum number of attractors that can be stored robustly) scales with network size, N, and you can derive statistical properties of the synaptic weights under assumptions about the optimal use of storage capacity. These properties can be compared to estimates from cortical measurements to gain insight into how memories might be stored in cortical circuits (Brunel 2016). In addition, a modified version of Hopfield networks is functionally equivalent to the attention mechanism used in transformer models, which are state-of-the-art machine learning models for language processing and other complex tasks (Ramsauer et al. 2020).

Exercise B.6.1. Simulate a Hopfield network with $N = 50$ neurons and use a random W satisfying the conditions $W_{jk} = W_{kj}$ and $W_{kk} = 0$. Plot $E(n)$ and verify that it is decreasing.

Exercise B.6.2. Train a Hopfield network on $M = 3$ MNIST digits (similar to figure B.13, but use equation (B.52) in place of equation (B.51)). Use three different digits for the three images (*e.g.*, do not choose two hand-drawn 2's). Then try running the Hopfield network using corrupted versions of the training images as initial conditions. Next try running the network with a *different* image representing one of the same digits (*e.g.*, if one of your training digits is a handwritten 2, then use a different handwritten 2 from the MNIST data set).

B.7 Training Readouts from Chaotic RNNs

The artificial neural network (ANN) model in section 4.2 in chapter 4 used *stationary* feedforward rate models to learn a mapping from *static* inputs, x, to *static* outputs, v. Specifically, the model and its inputs and outputs did not depend on time. Neural circuits are recurrent, produce dynamical (*i.e.*, time-varying) activity, and can learn to produce time-varying responses (*i.e.*, outputs) to time-varying stimuli (*i.e.*, inputs). The recurrent dynamical rate network models discussed in section 3.3 can be interpreted as recurrent ANNs. Unfortunately, learning the recurrent weight matrix in recurrent ANNs is more difficult, and it is not clear how it can be achieved with synaptic plasticity (Lillicrap and Santoro 2019). In this section, we describe a simpler method for training recurrent rate network models in which only a linear *readout* from the recurrent network is trained.

The recurrent network model is given as follows:

Recurrent neural network (RNN) model

$$\tau \frac{dr}{dt} = -r + f(Wr + X + Qz)$$

$$z = Rr.$$

(B.53)

Similar to the rate network model in equation (3.12) from section 3.3, $r(t) \in \mathbb{R}^N$ is vector of firing rates, $\tau > 0$ is a time constant, f is an f-I curve, and $X(t)$ is external input. An alternative formulation of RNN models (Sussillo and Abbott 2009) starts from the formulation of rate networks from equation (3.13) instead of equation (3.12), but the overall idea is the same.

Unlike equation (3.12), we have an additional term $z(t) \in \mathbb{R}^M$, which is a *readout* from the network, which is interpreted as output from the network. The goal is to find an R that makes $z(t)$ match a target time series, $y(t)$. The matrix $R \in \mathbb{R}^{M \times N}$ is a readout matrix, and $Q \in \mathbb{R}^{N \times M}$ is a feedback matrix that injects the readout back into the network.

The presence of feedback from the readout, $z(t)$, is the only technical difference between equations (B.53) and (3.12), but we will also use very different parameters. In section 3.3, we considered rate network models with just two populations, representing the average excitatory and inhibitory firing rates. For equation (B.53), we will consider a much larger network. In general, we should use $N \geq 100$ for things to work well. For the example considered here, we choose $N = 200$. Firing rates are also interpreted more abstractly. Instead of interpreting $r(t)$ as a vector of literal firing rates in physical units like hertz, we assume that rates take values in the interval $[-1, 1]$, meaning that they can even be negative. To achieve this, we use the hyperbolic tangent as an f-I curve:

$$f(x) = \tanh(x) = \frac{e^{2x} - 1}{e^{2x} + 1}.$$

This is a sigmoidal function with horizontal asymptotes at 1 and -1, respectively (figure B.14b). While allowing negative rates might seem odd, the idea is that more realistic rate values can be rescaled and shifted to lie in $[-1, 1]$, so it shouldn't matter. Moreover, this model is more abstract and the rates here might not be intended to represent literal firing rates, but are just a proxy for neural activity in general. We similarly allow entries in the connectivity matrices to take on positive and negative values without obeying Dale's law. Entries of the recurrent connectivity matrix, W, are drawn i.i.d. from a normal distribution with mean $\mu = E[W_{jk}] = 0$ and standard deviation $\sigma = \sqrt{\text{var}(W_{jk})} = \rho/\sqrt{N}$. In PyTorch, the matrix is generated by

```
W=rho*np.random.randn(N,N)/np.sqrt(N)
```

To better understand the dynamics of the network, let's first analyze the network without feedback, $Q = 0$, and without input, $X = 0$. In the absence of feedback, the model is equivalent to the rate network model from equation (3.12). Since $f(0) = \tanh(0) = 0$, there is a fixed

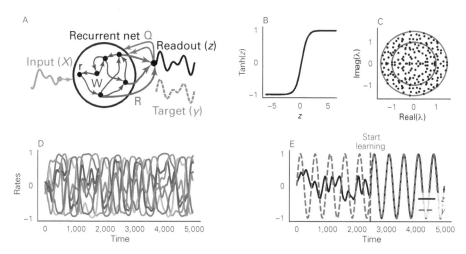

Figure B.14
Training readouts from a chaotic RNN. (A) Diagram of an RNN with readout trained to produce a target time series. (B) Hyperbolic tangent function used as an f-I curve. (C) Eigenvalues of W; unit circle shown in red, and circle of radius $\rho = 1.5$ shown in blue. (D) A sample of seven firing rates out of $N = 200$ from a network simulation without external input, $X(t) = 0$. (E) Output (z) and target (y). Learning and feedback were enabled only after time $t = 2,500$. The target is a one-dimensional ($M = 1$) sine wave. The code to reproduce this figure can be found in RNN.ipynb.

point at $r = 0$. Moreover, since $f'(0) = \tanh'(0) = 1$, the corresponding Jacobian matrix is

$$J = \frac{1}{\tau}[-I + W]$$

where I is the identity matrix. It is not difficult to check that the eigenvalues of J satisfy

$$\Lambda(J) = \frac{1}{\tau}[-1 + \Lambda(W)] \tag{B.54}$$

where $\Lambda(W)$ are the eigenvalues of W. Therefore, the fixed point is stable whenever all eigenvalues of W have a real part less than 1:

$$\mathrm{Re}(\Lambda(W)) < 1.$$

But what are the eigenvalues of the large, random matrix like W? Fortunately, *Girko's circular law* gives us an answer. This law says that if W is a random $N \times N$ matrix with i.i.d. entries satisfying $E[W_{jk}] = 0$, then (under some conditions) its eigenvalues are approximately uniformly distributed inside a circle centered at the origin in the complex plane. The radius of the smallest circle containing all of the eigenvalues is called the matrix's *spectral radius*, and Girko's circular law says that it is approximated by (Ginibre 1965; Girko 1985; Rajan and Abbott 2006)

$$r \approx \sqrt{N\mathrm{var}(W_{jk})}$$

where $\mathrm{var}(W_{jk})$ is the variance of the entries of W. A precise statement of the theorem requires more background in probability theory than is appropriate for this book. This approximation to the eigenvalues becomes increasingly accurate as $N \to \infty$. Note that for our model, $E[W_{jk}] = 0$ and $\mathrm{var}(W_{jk}) = \rho^2/N$, so we have

$$r \approx \rho.$$

Figure B.14c shows the eigenvalues of W when $N = 200$ and $\rho = 1.5$. The red circle has radius 1, and the blue circle has radius $\rho = 1.5$. Note that the eigenvalues appear approximately uniformly distributed within a circle of radius $\rho = 1.5$.

From equation (B.54), we can conclude that the eigenvalues of J lie in a circle of radius r/τ centered at $-1/\tau$, where r is the spectral radius of W. Hence, stability is likely whenever $\rho < 1$ and instability is likely when $\rho > 1$. Another way to visualize this result is that stability requires that the eigenvalues of W have a real part less than 1. You can see in figure B.14c that this requires the spectral radius, r, to be less than 1. Randomness in the eigenvalues can make the true spectral radius different from ρ, but when N is large, the transition between stability and instability is very likely to occur very close to $\rho = 1$.

In summary, if ρ is sufficiently smaller than 1, the fixed point at $\boldsymbol{r} = 0$ is stable and $\boldsymbol{r}(t) \to$ 0 as t increases. When ρ is sufficiently larger than 1, the fixed point is unstable. What kind of dynamics are produced by this instability? Note that $r_j(t) \in [0, 1]$, so the rates can't blow up. Instead, stability is lost through a high-dimensional bifurcation that produces intricate, chaotic dynamics (Sompolinsky, Crisanti, and Sommers 1988; Rajan and Abbott 2006). Essentially, each rate $r_j(t)$ traces out a complicated but relatively smooth time series. The magnitudes of τ and ρ control the timescale and disorderliness of the dynamics, respectively. The first half of figure B.14d (up to time 2,500) shows a sample of firing rates when $\rho = 1.5$ with feedback turned off.

The idea behind the RNN models studied in this section is that the dynamics of $\boldsymbol{r}(t)$ are rich and high-dimensional, so we should be able to mine them to produce an arbitrary time series. In other words, there should be a readout matrix, R, for which $\boldsymbol{z}(t) = R\boldsymbol{r}(t)$ matches our target time series, $\boldsymbol{y}(t)$. We can quantify the error of the network by the Euclidean distance between $\boldsymbol{z}(t)$ and $\boldsymbol{y}(t)$:

$$e = \|\boldsymbol{z} - \boldsymbol{y}\|^2 = \sum_{j=1}^{M}(z_j - y_j)^2.$$

We'd like to find an update to R that reduces e. Let's consider what happens when we change R by a small amount to get a new matrix. Consider an update to R of the form

$$R' = R + \epsilon \Delta R$$

where $\epsilon > 0$ is small; that is, we will consider the $\epsilon \to 0$ limit. The change to R causes a change to the readout, which we can write as

$$z' = z + \epsilon \Delta z + \mathcal{O}(\epsilon^2)$$

where $\mathcal{O}(\epsilon^2)$ are terms that go to zero, like ϵ^2 as $\epsilon \to 0$. Since $z = Rr$ and $z' = R'r$, we have

$$\Delta z = \lim_{\epsilon \to 0} \frac{z' - z}{\epsilon} = \lim_{\epsilon \to 0} \frac{R'r - Rr}{\epsilon} = \Delta Rr. \tag{B.55}$$

This calculation implicitly assumes that changing R does not affect $\boldsymbol{r}$, which is a valid assumption if we measure z' just after changing R, so feedback does not have enough time to change r by more than $\mathcal{O}(\epsilon^2)$. Now define the modified error:

$$e' = e + \epsilon\Delta e + \mathcal{O}(\epsilon^2).$$

We can compute the change to the error in a similar way:

$$
\begin{aligned}
\Delta e &= \lim_{\epsilon \to 0} \frac{e' - e}{\epsilon} \\
&= \lim_{\epsilon \to 0} \frac{1}{\epsilon} \left(\|z' - y\|^2 - \|z - y\|^2 \right) \\
&= \lim_{\epsilon \to 0} \frac{1}{\epsilon} \left((z + \epsilon\Delta z - y)^T (z + \epsilon\Delta z - y) - (z - y)^T (z - y) \right) \\
&= \lim_{\epsilon \to 0} \frac{1}{\epsilon} \left(\epsilon(z - y)^T \Delta z + \epsilon\Delta z^T (z - y) + \epsilon^2 \Delta z^T \Delta z \right) \\
&= 2(z - y)^T \Delta z
\end{aligned}
$$

where the last step follows from the fact that $(z - y)^T \Delta z$ is a scalar, so it is equal to its transpose, $\Delta z^T (z - y)$. Combining this result with equation (B.55), we have

$$\Delta e = 2(z - y)^T \Delta R r. \tag{B.56}$$

To decrease the error, we want to make sure that $\Delta e < 0$. To achieve this, we can use an update of the form $\Delta R = -(z - y)r^T$. Altogether, we can write the update as follows:

LMS update rule for readout matrices

$$R_{new} = R_{old} - \epsilon(z - y)r^T \tag{B.57}$$

---

where $\epsilon > 0$ is a small learning rate. If we use equation (B.57) to update R, then the error will decrease, so long as $\epsilon > 0$ is sufficiently small, $e \neq 0$, and $\|r\| \neq 0$. Note that the least mean squares (LMS) rule is a special case of the Delta rule from equation (4.7) in chapter 4, with $f'(z) = 1$, since the readout itself can be viewed as a single-layer ANN with weight matrix R and activation function $f(z) = z$.

The second half of figure B.14c,d (after time 2,500) shows the firing rates, readout, and targets after feedback and learning are turned on (using the LMS rule to update R). The readout quickly converges to the target.

The practice of training an RNN by updating only a set of readout weights is called *reservoir computing*, and the corresponding network model is called an *echo state network* or *liquid state machine* (Jaeger 2001; Maass, Natschläger, and Markram 2002; Jaeger 2004; Lukoševičius and Jaeger 2009; Sussillo and Abbott 2009). There are many variants of reservoir-computing algorithms. For example, some models do not use feedback ($Q = 0$), but feedback tends to improve learning. Some models include an input, $X(t)$, and use a target that depends on the input so the network needs to learn a mapping from an input time series, $X(t)$, to an output time series, $z(t)$. Improved supervised learning rules have been developed (Sussillo and Abbott 2009), in addition to "reward-modulated" learning rules that do not require knowledge $y(t)$, but only need $e(t) = \|z(t) - y(t)\|^2$ to update R (Hoerzer, Legenstein, and Maass 2014; Pyle and Rosenbaum 2019; Murray and Escola 2020).

Exercises B.7.1 and B.7.2 identify some weaknesses of the LMS learning rule as it is defined here and provide some approaches to improve it.

While the RNN models and learning rules considered here are far removed from biology, the dynamics of $r(t)$ in a trained RNN share statistical features of neural activity recorded in animals performing similar motor tasks and therefore can be used as abstract models of neural activity during motor tasks (Mante et al. 2013; Sussillo 2014; Sussillo et al. 2015; Pyle and Rosenbaum 2019; Murray and Escola 2020).

Technically, the feedback matrix, R, represents a recurrent connectivity matrix (at least when $Q \neq 0$) because there is a recurrent loop from r to z and back to r (Sussillo and Abbott 2009). However, most of the recurrence in the network is contained in the matrix, W, which is not trained by the learning rules considered in this section. RNNs can be trained by learning W, but the learning rules are more complicated, and it is not clear how they might be implemented in the brain (Lillicrap and Santoro 2019).

Exercise B.7.1. Use equation (B.56) to prove $\Delta e < 0$ for the LMS update rule in equation (B.57) when $e \neq 0$ and $\|r\| \neq 0$.

Hint: Note that $v^T v = \|v\|^2 > 0$ for any nonzero vector v.

Exercise B.7.2. In figure B.14, we compared the feedback and target only while learning was turned on. Ideally, the network would produce small errors even after learning is turned off (R held fixed after learning). Repeat the simulation in figure B.14, but turn off learning by holding R fixed after some time (keep the feedback turned on). You will see that the output does not match the target after learning is turned off. This is because our derivation of ΔR only required that the error is smaller immediately after updating R, but it will not necessarily keep the error small if R is not updated again on the next time step. In essence, with the LMS learning rule, the network doesn't really *learn* to produce the target, but the readout just chases the target around (Sussillo and Abbott 2009). Sussillo and Abbott 2009 showed that more stable learning can be achieved by modified learning rules called "FORCE learning" (where FORCE stands for "first-order reduced and controlled error").

B.8   DNNs and Backpropagation

In section 4.2 in chapter 4, we considered single-layered ANNs and how to train them with gradient descent. We now extend this discussion to multilayered ANNs, also called *deep neural networks (DNNs)*, which are defined by equations of the form

$$x = v^0$$
$$v^1 = f_1(W^1 v^0)$$
$$\dots$$
$$v = v^L = f_L(W^L v^{L-1}).$$

(B.58)

In other words,

$$v^\ell = f_\ell(W^\ell v^{\ell-1}), \quad \ell = 1, \dots, L$$

for $\ell = 1, \dots, L$ where $v^0 = x$ is the input to the network and $v^L = v$ is the output from the network. The vectors v^ℓ for $0 < \ell < L$ are called hidden states or *hidden layers*. Each v^ℓ is called the *activation* of layer ℓ. The function $f_\ell : \mathbb{R} \to \mathbb{R}$ is applied pointwise, and it is called the *activation function* for layer ℓ. The number, L, of layers is called the *depth* of the network. Each $v^\ell \in \mathbb{R}^{n_\ell}$ is a vector and its dimension, n_ℓ, is called the *width* of layer ℓ. Note that each W^ℓ is an $n_\ell \times n_{\ell-1}$ matrix. It is useful to define the preactivations:

$$z^\ell = W^\ell v^{\ell-1}$$

that satisfy $v^\ell = f(z^\ell)$.

DNNs are powerful algorithms. Indeed, *universal approximation theorem* says (roughly) that a DNN with just one hidden layer (*i.e.*, with $L = 2$) can approximate any reasonable function from x to v to any desired degree of accuracy (Hornik, Stinchcombe, and White 1989). However, the theorem does not tell us how wide the hidden layer needs to be or how we can learn the weights that approximate the desired function, so the theorem is less useful from a practical perspective.

We can again considered a supervised learning task with training data, $\{x^i, y^i\}_{i=1}^m$ and a cost function of the form

$$J(\theta) = \frac{1}{m} \sum_{i=1}^m L(v^i, y^i)$$

where $\theta = \{W^\ell\}_{\ell=1}^L$ is the collection of all parameters—that is, all weight matrices—and $L(v, y)$ is a loss function. And we can again learn through gradient descent on L for online gradient descent, or on J for full batch gradient descent. We will focus on online gradient descent here, but the calculations are analogous for full batch or stochastic gradient descent (SGD).

We now need to update all L matrices, W^ℓ, on each gradient descent step. The gradients of the loss with respect to W^ℓ are a little bit more difficult to derive for DNNs than for single-layer networks. We will not give a detailed derivation here, but we will outline the basic equations and result. The update to W^ℓ can be written as

$$\Delta W^\ell = -\epsilon \nabla_{W^\ell} L = -\epsilon e^\ell \left[v^{\ell-1}\right]^T \tag{B.59}$$

where $\epsilon > 0$ is a learning rate and

$$e^\ell = \nabla_{z^\ell} L$$

is the gradient of the loss with respect to the preactivation, z^ℓ. We will call e^ℓ the *error* of layer ℓ. The last layer's error is given by

$$e^L = [\nabla_v L] \circ f_L'(z^L). \tag{B.60}$$

The error terms from earlier layers can then be defined in terms of the layer after them:

$$e^\ell = \left[B^\ell e^{\ell+1}\right] \circ f_\ell'(z^\ell) \tag{B.61}$$

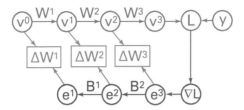

Figure B.15
A diagram of forward and backward passes for training a DNN with backpropagation. A forward pass (red pathways) computes the activations and loss according to equation (B.58). Then a backward pass (blue pathways) computes gradients and errors according to equation (B.62). The activations and errors are combined to compute the updates, ΔW^ℓ, to the connection matrices according to equation (B.59). The backward pass is sometimes interpreted as a separate feedforward network.

where

$$B^\ell = \left[W^{\ell+1} \right]^T$$

is the transpose of the weight matrix from the next layer.

Equations (B.87) through (B.89) can be used to compute the errors and gradient descent updates by working our way from the last layer backward to the first layer of the network. Specifically, after we compute the output and loss of the network, we can use equation (B.60) to compute e^L. We can then use equation (B.61) to compute e^{L-1}, e^{L-2}, and continue on, working our way backward to e^1. As we go, we can compute the weight updates of each layer using equation (B.59). This procedure is known as *backpropagation*.

Backpropagation can be viewed as propagating the error terms, e^ℓ, backward through the network. In particular, equation (B.61) represents a network in which layer $\ell + 1$ connects to layer ℓ with the connection matrix $B^\ell = \left[W^{\ell+1} \right]^T$. Specifically, we have

$$e^L = [\nabla_v L] \circ f'_\ell(z^\ell)$$

$$e^{L-1} = \left[B^{L-1} e^L \right] \circ f'_{L-1}(z^{L-1})$$

$$\cdots$$

$$e^1 = \left[B^1 e^2 \right] \circ f'_1(z^1).$$

(B.62)

This is similar to the "forward" network in equation (B.58) except that information flows backward through the network, we use the derivative of the activation function, and we multiply the activation function by the matrix product.

When training a network using backpropagation, we first apply equation (B.58) to compute activations, and then we compute the loss. This is called a *forward pass* through the network. We then use equation (B.62) to compute the errors and use equation (B.59) to compute the weight updates. This is called a *backward pass* through the network. These ideas are illustrated in figure B.15.

Backpropagation is an efficient way to compute gradients in DNNs, and gradient-based learning is a highly effective method for training these networks to perform difficult tasks. Partly for these reasons, many computational neuroscientists have tried to understand how backpropagation might be implemented or approximated in the brain (Lillicrap and Santoro

2019; Whittington and Bogacz 2019; Lillicrap et al. 2020). The basic idea behind many of these approaches is that equations (B.58) and (B.62) are like feedforward rate networks (in opposite directions) and equation (B.59) is similar to a plasticity rule. There are many difficulties with the direct interpretation of backpropagation in terms of neural circuits, of course.

One interpretation is that the backward pathway represents separate populations of neurons from the forward pathway (*i.e.*, e^ℓ represent a separate population of neurons from v^ℓ). In this case, the update, ΔW_{jk}^ℓ, to the weight that connects $v_k^{\ell-1}$ to v_j^ℓ depends on $e_k^{\ell-1}$. Plasticity rules in the brain are often believed to be largely *local*, meaning that an update to a synaptic weight should be a function of the activity of the neurons that the weight connects. In other words, ΔW_{jk}^ℓ should be a function of $v_k^{\ell-1}$ and v_j^ℓ. Hence, backpropagation is not a local plasticity rule under this interpretation.

An alternative interpretation is that v_j^ℓ and e_j^ℓ represents the same neuron during two phases, an inference phase during which information flows forward in the network to compute v^ℓ, and a learning phase during which information flows backward through the same network to compute e^ℓ. This idea is appealing because there are feedback pathways in the cortex (*i.e.*, V1 connects to V2 and V2 connects back to V1). One problem with this interpretation is that it is not clear how the brain would distinguish between activity during the two phases.

A major problem with both interpretations given here is that the backward connectivity matrices need to be equal to the transpose of the forward connectivity matrices, $B^\ell = \left[W^{\ell+1}\right]^T$, for backpropagation. This requirement that is sometimes called *weight transport*. There is no strong evidence of weight transport in cortical circuits, and it is not clear how it could be enforced, although some plasticity rules can help enforce it approximately (Kolen and Pollack 1994; Akrout et al. 2019; Shervani-Tabar and Rosenbaum 2022).

A third interpretation that was posed more recently is that v_j^ℓ and e_j^ℓ are represented by the same neuron, but in different ways (Naud and Sprekeler 2018; Richards and Lillicrap 2019; Payeur et al. 2021). Calcium channels and some other mechanism can cause neurons to emit short bursts of spikes. Under this third interpretation, neurons separately encode v^ℓ and e^ℓ in the rate of spikes and the rate of bursts, respectively. Through a combination of architecture, short-term synaptic plasticity, and burst-dependent long term synaptic plasticity, this multiplexed encoding of v^ℓ and e^ℓ can approximate backpropagation. While this interpretation is more complicated than the other two interpretations mentioned here, it is consistent with many features of cortical circuits. Cortical circuits are complicated, so a more complicated explanation of learning should not necessarily be ruled out.

Exercise B.8.1. Extend the network from `SingleLayerANN.ipynb` to $L = 3$ layers, and train the weights using backpropagation.

References

Abbott, L. F. 1999. "Lapicque's Introduction of the Integrate-and-Fire Model Neuron (1907)." *Brain Research Bulletin* 50(5–6): 303–304.

Adesnik, H., W. Bruns, H. Taniguchi, Z. J. Huang, and M. Scanziani. 2012. "A Neural Circuit for Spatial Summation in Visual Cortex." *Nature* 490(7419): 226–31.

Ahmadian, Y., and K. D. Miller. 2021. "What is the Dynamical Regime of Cerebral Cortex?" *Neuron* 109(21): 3373–3391.

Akrout, M., C. Wilson, P. Humphreys, T. Lillicrap, and D. B. Tweed. 2019. "Deep Learning without Weight Transport." *Advances in Neural Information Processing Systems 32*, arXiv: 1904.05391.

Allman, J., F. Miezin, and E. McGuinness. 1985. "Stimulus Specific Responses from beyond the Classical Receptive Field: Neurophysiological Mechanisms for Local-Global Comparisons in Visual Neurons." *Annual Review of Neuroscience* 8: 407–430.

Amit, D. J., H. Gutfreund, and H. Sompolinsky. 1985. "Spin-Glass Models of Neural Networks." *Physical Review A* 32(2): 1007.

Amit, D. J., and M. Tsodyks. 1991. "Quantitative Study of Attractor Neural Network Retrieving at Low Spike Rates. I. Substrate-Spikes, Rates and Neuronal Gain." *Network: Computation in Neural Systems* 2(3): 259.

Averbeck, B. B., P. E. Latham, and A. Pouget. 2006. "Neural Correlations, Population Coding and Computation." *Nature Reviews Neuroscience* 7(5): 358–366.

Averbeck, B. B., and D. Lee. 2006. "Effects of Noise Correlations on Information Encoding and Decoding." *Journal of Neurophysiology* 95(6): 3633–3644.

Badel, L., S. Lefort, T. K. Berger, C. C. Petersen, W. Gerstner, and M. J. Richardson. 2008. "Extracting Non-linear Integrate-and-Fire Models from Experimental Data Using Dynamic I–V Curves." *Biological Cybernetics* 99(4): 361–370.

Bastos, A. M., W. M. Usrey, R. A. Adams, G. R. Mangun, P. Fries, and K. J. Friston. 2012. "Canonical Microcircuits for Predictive Coding." *Neuron* 76(4): 695–711.

Bellec, G., F. Scherr, A. Subramoney, E. Hajek, D. Salaj, R. Legenstein, and W. Maass. 2020. "A Solution to the Learning Dilemma for Recurrent Networks of Spiking Neurons." *Nature Communications* 11(1): 1–15.

Brette, R. 2019. "Is Coding a Relevant Metaphor for the Brain?" *Behavioral and Brain Sciences* 42.

Brette, R., and W. Gerstner. 2005. "Adaptive Exponential Integrate-and-Fire Model as an Effective Description of Neuronal Activity." *Journal of Neurophysiology* 94(5): 3637–3642.

Brunel, N. 2016. "Is Cortical Connectivity Optimized for Storing Information?" *Nature Neuroscience* 19(5): 749–755.

Brunel, N., and M. C. Van Rossum. 2007. "Lapicque's 1907 Paper: From Frogs to Integrate-and-Fire." *Biological Cybernetics* 97(5): 337–339.

Brunel, N., and X.-J. Wang. 2003. "What Determines the Frequency of Fast Network Oscillations with Irregular Neural Discharges? I. Synaptic Dynamics and Excitation-Inhibition Balance." *Journal of Neurophysiology* 90(1): 415–430.

Canavier, C. C., and S. Achuthan. 2010. "Pulse Coupled Oscillators and the Phase Resetting Curve." *Mathematical Biosciences* 226(2): 77–96.

Capogna, M., P. E. Castillo, and A. Maffei. 2021. "The Ins and Outs of Inhibitory Synaptic Plasticity: Neuron Types, Molecular Mechanisms and Functional Roles." *European Journal of Neuroscience* 54(8): 6882–6901.

Cardin, J. A., M. Carlén, K. Meletis, U. Knoblich, F. Zhang, K. Deisseroth, L.-H. Tsai, and C. I. Moore. 2009. "Driving Fast-Spiking Cells Induces Gamma Rhythm and Controls Sensory Responses." *Nature* 459(7247): 663–667.

Castillo, P. E., C. Q. Chiu, and R. C. Carroll. 2011. "Long-Term Plasticity at Inhibitory Synapses." *Current Opinion in Neurobiology* 21(2): 328–338.

Churchland, M. M., M. Y. Byron, J. P. Cunningham, et al. 2010. "Stimulus Onset Quenches Neural Variability: A Widespread Cortical Phenomenon." *Nature Neuroscience* 13(3): 369–378.

Cornford, J., D. Kalajdzievski, M. Leite, A. Lamarquette, D. M. Kullmann, and B. A. Richards. 2020. "Learning to Live with Dale's Principle: ANNs with Separate Excitatory and Inhibitory Units." *International Conference on Learning Representations*.

Curto, C., A. Degeratu, and V. Itskov. 2013. "Encoding Binary Neural Codes in Networks of Threshold-Linear Neurons." *Neural Computation* 25(11): 2858–2903.

Dayan, P., and L. F. Abbott. 2001. *Theoretical Neuroscience*. Cambridge, MA: MIT Press.

Ermentrout, G. B., and N. Kopell. 1986. "Parabolic Bursting in an Excitable System Coupled with a Slow Oscillation." *SIAM Journal on Applied Mathematics* 46(2): 233–253.

Ermentrout, G. B., and D. H. Terman. 2010. *Mathematical Foundations of Neuroscience*. New York: Springer Science & Business Media.

Fourcaud–Trocme, N., D. Hansel, C. van Vreeswijk, and N. Brunel. 2003. "How Spike Generation Mechanisms Determine the Neuronal Response to Fluctuating Inputs." *Journal of Neuroscience* 23: 11628–11640.

Gerstein, G. L., and B. Mandelbrot. 1964. "Random Walk Models for the Spike Activity of a Single Neuron." *Biophysical Journal* 4(1): 41–68.

Gerstner, W., W. M. Kistler, R. Naud, and L. Paninski. 2014. *Neuronal Dynamics: From Single Neurons to Networks and Models of Cognition*. Cambridge: Cambridge University Press.

Ginibre, J. 1965. "Statistical Ensembles of Complex, Quaternion, and Real Matrices." *Journal of Mathematical Physics* 6(3): 440–449.

Ginzburg, I., and H. Sompolinsky. 1994. "Theory of Correlations in Stochastic Neural Networks." *Physical Review E* 50(4): 3171.

Girko, V. L. 1985. "Circular Law." *Theory of Probability & Its Applications* 29(4): 694–706.

Guerguiev, J., T. P. Lillicrap, and B. A. Richards. 2017. "Towards Deep Learning with Segregated Dendrites." *eLife* 6: e22901.

Hebb, D. O. 1949. *The Organization of Behavior: A Neuropsychological Theory*. New York: John Wiley & Sons.

Hennequin, G., E. J. Agnes, and T. P. Vogels. 2017. "Inhibitory Plasticity: Balance, Control, and Codependence." *Annual Review of Neuroscience* 40(1): 557–579.

Hirsch, M. W., S. Smale, and R. L. Devaney. *Differential Equations, Dynamical Systems, and an Introduction to Chaos*. Waltham, MA: Academic Press, 2012.

Hodgkin, A. L., and A. F. Huxley. 1952. "A Quantitative Description of Membrane Current and Its Application to Conduction and Excitation in Nerve." *Journal of Physiology* 117(4): 500.

Hoerzer, G. M., R. Legenstein, and W. Maass. 2014. "Emergence of Complex Computational Structures from Chaotic Neural Networks through Reward-Modulated Hebbian Learning." *Cerebral Cortex* 24(3): 677–690.

Hopfield, J. J. 1982. "Neural Networks and Physical Systems with Emergent Collective Computational Abilities." *Proceedings of the National Academy of Sciences* 79(8): 2554–2558.

Hoppensteadt, F. C., and E. M. Izhikevich. 1997. *Weakly Connected Neural Networks*. New York: Springer Science & Business Media.

Hornik, K., M. Stinchcombe, and H. White. 1989. "Multilayer Feedforward Networks Are Universal Approximators." *Neural Networks* 2(5): 359–366.

Huang, C., D. Ruff, R. Pyle, R. Rosenbaum, M. Cohen, and B. Doiron. 2019. "Circuit Models of Low-Dimensional Shared Variability in Cortical Networks." *Neuron* 101(2): 337–348.

Huh, D., and T. J. Sejnowski. 2018. "Gradient Descent for Spiking Neural Networks." *Advances in Neural Information Processing Systems* 31.

Izhikevich, E. M. 2003. "Simple Model of Spiking Neurons." *IEEE Transactions on Neural Networks* 14(6): 1569–1572.

Izhikevich, E. M. 2004. "Which Model to Use for Cortical Spiking Neurons?" *IEEE Transactions on Neural Networks* 15(5): 1063–1070.

Izhikevich, E. M. 2007. *Dynamical Systems in Neuroscience*. Cambridge, MA: MIT press.

Jaeger, H. 2001. *The "Echo State" Approach to Analysing and Training Recurrent Neural Networks-with an Erratum Note*. GMD Technical Report. Bonn: German National Research Center for Information Technology.

Jaeger, H. 2004. "Harnessing Nonlinearity: Predicting Chaotic." *Science* 304(5667): 78–80. https://www.science.org/doi/10.1126/science.1091277.

Jolivet, R., T. J. Lewis, and W. Gerstner. 2004. "Generalized Integrate-and-Fire Models of Neuronal Activity Approximate Spike Trains of a Detailed Model to a High Degree of Accuracy." *Journal of Neurophysiology* 92(2): 959–976.

Jolivet, R., F. Schürmann, T. K. Berger, R. Naud, W. Gerstner, and A. Roth. 2008. "The Quantitative Single-Neuron Modeling Competition." *Biological Cybernetics* 99(4–5): 417.

Kell, A. J., D. L. Yamins, E. N. Shook, S. V. Norman-Haignere, and J. H. McDermott. 2018. "A Task-Optimized Neural Network Replicates Human Auditory Behavior, Predicts Brain Responses, and Reveals a Cortical Processing Hierarchy." *Neuron* 98(3): 630–644.

Khaligh-Razavi, S.-M., and N. Kriegeskorte. 2014. "Deep Supervised, But Not Unsupervised, Models May Explain IT Cortical Representation." *PLoS Computational Biology* 10(11): e1003915.

Knight, B. W. 1972. "Dynamics of Encoding in a Population of Neurons." *Journal of General Physiology* 59(6): 734–766.

Kolen, J. F., and J. B. Pollack. 1994. "Backpropagation without Weight Transport." *Proceedings of 1994 IEEE International Conference on Neural Networks (ICNN'94)*, 3: 1375–1380.

Kopell, N., and G. Ermentrout. 1990. "Phase Transitions and Other Phenomena in Chains of Coupled Oscillators." *SIAM Journal on Applied Mathematics* 50(4): 1014–1052.

Kopell, N., G. B. Ermentrout, M. Whittington, and R. D. Traub. 2000. "Gamma Rhythms and Beta Rhythms Have Different Synchronization Properties." *Proceedings of the National Academy of Sciences* 97(4): 1867–1872.

Lapique, L. 1907. "Recherches Quantitatives sur l'Excitation Electrique des Nerfs Traitee Comme Une Polarization." *Journal of Physiology and Pathololgy* 9: 620–635.

LeCun, Y., L. Bottou, Y. Bengio, and P. Haffner. 1998. "Gradient-Based Learning Applied to Document Recognition." *Proceedings of the IEEE* 86(11): 2278–2324.

Li, Y., Y. Guo, S. Zhang, S. Deng, Y. Hai, and S. Gu. 2021. "Differentiable Spike: Rethinking Gradient-Descent for Training Spiking Neural Networks." *Advances in Neural Information Processing Systems* 34.

Lillicrap, T. P., and A. Santoro. 2019. "Backpropagation through Time and the Brain." *Current Opinion in Neurobiology* 55:82–89.

Lillicrap, T. P., A. Santoro, L. Marris, C. J. Akerman, and G. Hinton. 2020. "Backpropagation and the Brain." *Nature Reviews Neuroscience* 21(6): 335–346.

Lindner, B., J. Garcıa-Ojalvo, A. Neiman, and L. Schimansky-Geier. 2004. "Effects of Noise in Excitable Systems." *Physics Reports* 392(6): 321–424.

Lindsay, G. 2021. *Models of the Mind: How Physics, Engineering and Mathematics Have Shaped Our Understanding of the Brain*. New York: Bloomsbury Publishing.

Lukoševičius, M., and H. Jaeger. 2009. "Reservoir Computing Approaches to Recurrent Neural Network Training." *Computer Science Review* 3(3): 127–149.

Luz, Y., and M. Shamir. 2012. "Balancing Feed-forward Excitation and Inhibition via Hebbian Inhibitory Synaptic Plasticity." *PLoS Computational Biology* 8(1): e1002334.

Maass, W., T. Natschläger, and H. Markram. 2002. "Real-time Computing without Stable States: A New Framework for Neural Computation Based on Perturbations." *Neural Computation* 14(11): 2531–2560.

Mante, V., D. Sussillo, K. V. Shenoy, and W. T. Newsome. 2013. "Context-Dependent Computation by Recurrent Dynamics in Prefrontal Cortex." *Nature* 503(74): 78–84.

McCulloch, W. S., and W. Pitts. 1943. "A Logical Calculus of the Ideas Immanent in Nervous Activity." *Bulletin of Mathematical Biophysics* 5(4): 115–133.

McLachlan, G. J. 2005. *Discriminant Analysis and Statistical Pattern Recognition.* New York: John Wiley & Sons.

Moreno-Bote, R., J. Beck, I. Kanitscheider, X. Pitkow, P. Latham, and A. Pouget. 2014. "Information-Limiting Correlations." *Nature Neuroscience* 17(10): 1410–1417.

Murray, J. M., and G. S. Escola. 2020. "Remembrance of Things Practiced with Fast and Slow Learning in Cortical and Subcortical Pathways." *Nature Communications* 11(1): 1–12.

Naud, R., and H. Sprekeler. 2018. "Sparse Bursts Optimize Information Transmission in a Multiplexed Neural Code." *Proceedings of the National Academy of Sciences* 115(27): E6329–E6338.

Neftci, E. O., H. Mostafa, and F. Zenke. 2019. "Surrogate Gradient Learning in Spiking Neural Networks: Bringing the Power of Gradient-Based Optimization to Spiking Neural Networks." *IEEE Signal Processing Magazine* 36(6): 51–63.

Okun, M., A. Naim, and I. Lampl. 2010. "The subthreshold Relation Between Cortical Local Field Potential and Neuronal Firing Unveiled by Intracellular Recordings in Awake Rats." *Journal of Neuroscience* 30(12): 4440–4448.

Oprisan, S., A. Prinz, and C. Canavier. 2004. "Phase Resetting and Phase Locking in Hybrid Circuits of One Model and One Biological Neuron." *Biophysical Journal* 87(4): 2283–2298.

Ozeki, H., I. M. Finn, E. S. Schaffer, K. D. Miller, and D. Ferster. 2009. "Inhibitory Stabilization of the Cortical Network Underlies Visual Surround Suppression." *Neuron* 62(4): 578–592.

Panzeri, S., M. Moroni, H. Safaai, and C. D. Harvey. 2022. "The Structures and Functions of Correlations in Neural Population Codes." *Nature Reviews Neuroscience,* 1–17.

Payeur, A., J. Guerguiev, F. Zenke, B. A. Richards, and R. Naud. 2021. "Burst-Dependent Synaptic Plasticity Can Coordinate Learning in Hierarchical Circuits." *Nature Neuroscience* 24(7): 1010–1019.

Perkel, D. H., J. H. Schulman, T. H. Bullock, G. P. Moore, and J. P. Segundo. 1964. "Pacemaker Neurons: Effects of Regularly Spaced Synaptic Input." *Science* 145(3627): 61–63.

Pfeffer, C. K., M. Xue, M. He, Z. J. Huang, and M. Scanziani. 2013. "Inhibition of Inhibition in Visual Cortex: The Logic of Connections Between Molecularly Distinct Interneurons." *Nature Neuroscience* 16(8): 1068–1076.

Pyle, R., and R. Rosenbaum. 2019. "A Reservoir Computing Model of Reward-Modulated Motor Learning and Automaticity." *Neural Computation* 31(7): 1430–1461.

Rajan, K., and L. F. Abbott. 2006. "Eigenvalue Spectra of Random Matrices for Neural Networks." *Physical Review Letters* 97(18): 188104.

Ramsauer, H., B. Schäfl, J. Lehner, et al. 2020. "Hopfield Networks is All You Need." *International Conference on Learning Representations.* https://openreview.net/forum?id=tL89RnzIiCd.

Renart, A., J. De La Rocha, P. Bartho, L. Hollender, N. Parga, A. Reyes, and K. D. Harris. 2010. "The Asynchronous State in Cortical Circuits." *Science* 327(5965): 587–590.

Ricciardi, L. M., and L. Sacerdote. 1979. "The Ornstein-Uhlenbeck Process as a Model for Neuronal Activity." *Biological Cybernetics* 35(1): 1–9.

Richards, B. A., and T. P. Lillicrap. 2019. "Dendritic Solutions to the Credit Assignment Problem." *Current Opinion in Neurobiology* 54: 28–36.

Richards, B. A., T. P. Lillicrap, P. Beaudoin, et al. 2019. "A Deep Learning Framework for Neuroscience." *Nature Neuroscience* 22(11): 1761–1770.

Richardson, M. J. 2007. "Firing-Rate Response of Linear and Nonlinear Integrate-and-Fire Neurons to Modulated Current-Based and Conductance-Based Synaptic Drive." *Physical Review E* 76(2): 021919.

Robinson, A. 2018. "Did Einstein Really Say That?" *Nature* 557(7703): 30–31.

Rosenbaum, R., and K. Josić. 2011. "Mechanisms That Modulate the Transfer of Spiking Correlations." *Neural Computation* 23(5): 1261–1305.

Sanzeni, A., B. Akitake, H. C. Goldbach, C. E. Leedy, N. Brunel, and M. H. Histed. 2020. "Inhibition Stabilization Is a Widespread Property of Cortical Networks." *elife* 9: e54875.

Schrimpf, M., J. Kubilius, M. J. Lee, N. A. R. Murty, R. Ajemian, and J. J. DiCarlo. 2020. "Integrative Benchmarking to Advance Neurally Mechanistic Models of Human Intelligence." *Neuron* 108(3): 413–423.

Schulz, A., C. Miehl, M. J. Berry II, and J. Gjorgjieva. 2021. "The Generation of Cortical Novelty Responses through Inhibitory Plasticity." *eLife* 10: e65309.

Shadlen, M. N., and W. T. Newsome. 1994. "Noise, Neural Codes and Cortical Organization." *Current Opinion in Neurobiology* 4(4): 569–579.

Shadlen, M. N., and W. T. Newsome. 1998. "The Variable Discharge of Cortical Neurons: Implications for Connectivity, Computation, and Information Coding." *Journal of Neuroscience* 18(10): 3870–3896.

Shervani-Tabar, N., and R. Rosenbaum. 2022. "Meta-Learning Biologically Plausible Plasticity Rules with Random Feedback Pathways," arXiv preprint arXiv: 2210.16414.

Smith, M. A., and A. Kohn. 2008. "Spatial and Temporal Scales of Neuronal Correlation in Primary Visual Cortex." *Journal of Neuroscience* 28(48): 12591–12603.

Softky, W. R., and C. Koch. 1993. "The Highly Irregular Firing of Cortical Cells is Inconsistent With Temporal Integration of Random EPSPs." *Journal of Neuroscience* 13(1): 334–350.

Sompolinsky, H., A. Crisanti, and H.-J. Sommers. 1988. "Chaos in Random Neural Networks." *Physical Review Letters* 61(3): 259.

Stiefel, K. M., and G. B. Ermentrout. 2016. "Neurons as Oscillators." *Journal of Neurophysiology* 116(6): 2950–2960.

Strogatz, S. H. 2018. *Nonlinear Dynamics and Chaos: With Applications to Physics, Biology, Chemistry, and Engineering.* Boca Raton, FL: CRC press.

Strogatz, S. H., and I. Stewart. 1993. "Coupled Oscillators and Biological Synchronization." *Scientific American* 269(6): 102–109.

Sussillo, D. 2014. "Neural Circuits as Computational Dynamical Systems." *Current Opinion in Neurobiology* 25: 156–163.

Sussillo, D., and L. F. Abbott. 2009. "Generating Coherent Patterns of Activity from Chaotic Neural Networks." *Neuron* 63(4): 544–557.

Sussillo, D., M. M. Churchland, M. T. Kaufman, and K. V. Shenoy. 2015. "A Neural Network That Finds a Naturalistic Solution for the Production of Muscle Activity." *Nature Neuroscience* 18(7): 1025–1033.

Tsodyks, M. V., W. E. Skaggs, T. J. Sejnowski, and B. L. McNaughton. 1997. "Paradoxical Effects of External Modulation of Inhibitory Interneurons." *Journal of Neuroscience* 17(11): 4382–4388.

Vogels, T. P., H. Sprekeler, F. Zenke, C. Clopath, and W. Gerstner. 2011. "Inhibitory Plasticity Balances Excitation and Inhibition in Sensory Pathways and Memory Networks." *Science* 334(6062): 1569–1573.

Vogels, T. P., R. C. Froemke, N. Doyon, et al. 2013. "Inhibitory Synaptic Plasticity: Spike Timing-Dependence and Putative Network Function." *Frontiers in Neural Circuits* 7(119).

Vreeswijk, C. van, and H. Sompolinsky. 1998. "Chaotic Balanced State in a Model of cortical Circuits." *Neural Computation* 10(6): 1321–1371.

Whittington, J. C., and R. Bogacz. 2019. "Theories of Error Backpropagation in the Brain." *Trends in Cognitive Sciences* 23(3): 235–250.

Whittington, M. A., R. D. Traub, N. Kopell, B. Ermentrout, and E. H. Buhl. 2000. "Inhibition-Based Rhythms: Experimental and Mathematical Observations on Network Dynamics." *International Journal of Psychophysiology* 38(3): 315–336.

Widrow, B., and M. E. Hoff. 1960. *Adaptive Switching Circuits.* Technical report. Stanford Electronics Labs, Stanford University, Stanford, CA.

Wilson, H. R., and J. D. Cowan. 1972. "Excitatory and Inhibitory Interactions in Localized Populations of Model Neurons." *Biophysical Journal* 12(1): 1–24.

Wilson, H. R., and J. D. Cowan. 1973. "A Mathematical Theory of the Functional Dynamics of Cortical and Thalamic Nervous Tissue." *Kybernetik* 13(2): 55–80.

Yamins, D. L., H. Hong, C. F. Cadieu, E. A. Solomon, D. Seibert, and J. J. DiCarlo. 2014. "Performance-Optimized Hierarchical Models Predict Neural Responses in Higher Visual Cortex." *Proceedings of the National Academy of Sciences* 111(23): 8619–8624.

Zohary, E., M. N. Shadlen, and W. T. Newsome. 1994. "Correlated Neuronal Discharge Rate and Its Implications for Psychophysical Performance." *Nature* 370(6485): 140–143.

Index

Computational Neuroscience

Terence J. Sejnowski and Tomaso A. Poggio, editors